DIRECTIONS IN DEVELOPMENT

Private Sector Participation in the Water Supply and Wastewater Sector

Lessons from Six Developing Countries

Daniel Rivera

The World Bank
Washington, D.C.

Manufactured in the United States of America
First printing September 1996

The findings, interpretations, and conclusions expressed in this study are entirely those of the authors and should not be attributed in any manner to the World Bank, to its affiliated organizations, or to the members of its Board of Executive Directors or the countries they represent.

Cover photograph: A potable water plant in Quito, Ecuador, by Willie Heinz/ Inter-American Development Bank.

Daniel Rivera is on the staff of the National Planning Department of Colombia (Calle 26, No. 13-19). When this study was written, he was a consultant both to the Transport, Water, and Urban Development Department in the Environmentally Sustainable Development Vice-presidency of the World Bank and to the Technical Department in the Latin America and the Caribbean Regional Office of the Bank.

Library of Congress Cataloging-in-Publication Data

Rivera, Daniel, 1958–
Private sector participation in the water supply and wastewater sector : lessons from six developing countries / Daniel Rivera.
p. cm. — (Directions in development)
Includes bibliographical references.
ISBN 0-8213-3681-9
1. Water supply—Developing countries—Planning—Case studies.
2. Sewage disposal—Developing countries—Planning—Case studies.
3. Privatization—Developing countries—Case studies. I. Title.
II. Series: Directions in development (Washington, D.C.)
HD1702.R58 1996
363.6'1'091724—dc20 96-33292
CIP

Contents

Boxes

Figures

Tables

Foreword

This study is the fruit of a joint effort between the National Planning Department of Colombia and the World Bank. This type of collaboration is a positive example of the great potential for coordinated work between the World Bank and its member countries in areas of common interest.

Clearly, the infrastructure challenge for developing countries will remain with us for some time to come. In the water supply and wastewater sector in particular, a significant amount of resources—along with reforms of policies and institutions—are urgently needed. When considering broader macroeconomic reforms, many countries have chosen partnerships between the public and private sectors as a way to expand service coverage and quality. There are other compelling reasons for this as well, including the need to generate resources to finance the large infrastructure investment required, to reduce the public fiscal burden, and to increase productivity.

The financial challenge in the water supply and wastewater sector is great. And the current tradeoff of risks and rewards poses special challenges to governments, financial institutions, and the private sector to develop innovative solutions. The task has proved to be more complex than initially presumed.

This study is meant to highlight the gains and the limitations of recent experiences with private sector participation in the water supply and wastewater sector in developing countries. It aims to draw out valuable lessons that will be useful for developing countries to consider when planning and implementing current and future projects linked to private sector participation.

The evidence of this study reveals that in most cases private sector participation led to expanded coverage, improved quality of services, and gains in management and improved customer service. However, the study findings also suggest that much remains to be done and that the sustainability of the reforms is not a foregone conclusion for several reasons. Private sector participation cannot compensate for institutional inefficiencies and public sector failures, nor is it realistic to expect significant private financing in the initial years when risks are the highest. This means that there is a need for realism in the expectation for private sector financing. The truth is that strengthening public administration will be an essential step in the reform process. Complementary reforms will be needed to consolidate initial gains and to address the burden of long-standing structural limitations, especially in the areas of financing, regulation, and pricing of water services.

In essence, relying on standardized prescriptions or models does not always lead to desired results, as amply documented here. This study also invites further analytical work on the array of technical, institutional, and financial issues confronting developing nations in their ongoing effort to craft innovative and practical solutions to improve essential services and thus the quality of life for the general population.

Sri-Ram Aiyer
Director, Technical Department
Latin America and the Caribbean
Regional Office
The World Bank

Anthony Pellegrini
Director, Transportation, Water, and Urban Development
Environmentally Sustainable Development Vice-presidency
The World Bank

Preface

Reform of the water and wastewater sector in developing countries began in the 1980s and is still under way. The assessment of countries' ongoing experiences with private sector participation in providing these services can offer valuable lessons to other cities and countries facing similar challenges and constraints. Appropriate adjustments are being made as these lessons emerge.

This study synthesizes the results of six recent experiences with private sector participation in the water and wastewater sectors of developing countries. Its objectives are threefold. The first is to evaluate the results of these six cases and to describe the challenges and opportunities faced by governments and by water utilities, public as well as private, to improve the quantity and the quality of services. The second objective is to assess whether local and national governments have taken full advantage of the potential benefits of private sector participation and to explore why some of those benefits may not be fully realized. The third objective is to draw lessons from these experiences that can improve the planning and implementation of these water and wastewater projects and others elsewhere in which the private sector will participate.

The six case studies were selected by analyzing various models of private sector participation in different regions of the world, under different types of contracts and administrative arrangements and subject to various regulatory schemes. The principal technical, financial, and regulatory features of each case are presented in the appendix. For one of the six principal cases (Guinea), as well as for four other cases described briefly in Chapter 2 (Corrientes and Tucumán, Argentina, and Puerto Vallarta and Aguascalientes, Mexico), material was derived from secondary sources. In particular, the evolution of the private sector participation process in Guinea was based on a recent study by Penelope Brook-Cowen (1996) and on an earlier study by Thelma Triche (1990).

Most of the private sector arrangements described here are in their initial stage of development. For that reason the study does not try to make definitive statements about the success or failure of any of them, but instead assesses the *direction* of the changes that have taken place to date.

Following the Introduction, Chapter 1 reviews the broader economic and institutional context within which private sector providers have participated in the water and wastewater sector in developing countries.

Chapter 2 follows with a profile of private sector arrangements in the six cases under study and a brief description of other arrangements elsewhere. The impact of these arrangements on service levels and quality as well as the cost of services to consumers are evaluated in Chapter 3, and the constraints to implementing more adequate water-pricing systems are discussed in Chapter 4. The performance of private firms and the efficiency gains achieved through their involvement are assessed in Chapter 5. Chapter 6 discusses the financial aspects of private sector participation, including the challenges and opportunities of financing the expansion of services. The regulation of private sector arrangements is analyzed in Chapter 7. Finally, Chapter 8 summarizes the lessons learned and offers recommendations for increasing the likelihood of success of existing and future private sector arrangements.

Acronyms and Abbreviations

ACUACAR	Aguas de Cartagena S.A. (Colombia)
AGBAR	Aguas de Barcelona (Spain)
AGUAKAN	Aguas de Cancún S.A. (México)
BOT	Build-operate-transfer
CAPA	Comisión de Agua Potable y Alcantarillado (Quintana Roo, México)
CNA	Comisión Nacional del Agua (Mexico)
CORFO	Corporación de Fomento de la Producción (Chile)
CRA	Comisión de Regulación de Agua Potable y Saneamiento (Colombia)
DHC	Desarrollos Hidráulicos de Cancún S.A. (México)
EBRD	European Bank for Reconstruction and Development
EIB	European Investment Bank
EMOS	Empresa Metropolitana de Obras Sanitarias S.A. (Santiago, Chile)
EPD	Empresas Públicas Distritales (Cartagena, Colombia)
ETOSS	Ente Tripartito de Obras y Servicios Sanitarios (Buenos Aires, Argentina)
GMD	Grupo Mexicano de Desarrollo
IDA	International Development Association
IDB	Inter-American Development Bank
IFC	International Finance Corporation
MIGA	Multilateral Investment Guarantee Agency
O&M	Operation and management
OSN	Obras Sanitarias de la Nación (Buenos Aires, Argentina)
PAN	Partido de Acción Nacional de México

SEAPAL	Servicios de Agua Potable y Alcantarillado (Puerto Vallarta, México)
SEEG	Société d'Exploitation des Eaux de Guinée
SNG	Saur-Neptun-Gdansk (Gdansk, Poland)
SOEs	State-owned enterprises
SONEG	Société Nationale des Eaux de Guinée
SSS	Superintendencia de Servicios Sanitarios (Chile)
UFW	Unaccounted-for water

Acknowledgments

I would like to express my warm thanks to Klas Ringskog and Guillermo Yepes for their invaluable insights and sustained commitment to this study, as well as their personal support. I also have gained considerably from discussions with and comments by Rubén Avendaño, John Briscoe, Penelope Brook-Cowen, Augusta Dianderas, Vincent Gouarne, Walter Stottmann, and Carlos Velez. Similarly, I want to thank Sri-Ram Aiyer and Anthony Pellegrini, who supported the idea of initiating this study.

I want to acknowledge the key role played by the National Planning Department of Colombia as well as the World Bank for the financial support, and express my sincere gratitude to all teams of public agencies and private companies who shared with me their time, concerns, and aspirations. I am also indebted to William Mayville for his valuable editorial work.

And I especially want to thank my wife, because in many ways this work is as much hers as it is mine.

Summary

In the water supply and wastewater sector, partnership between the private and the public sectors has emerged as a promising way to improve the performance of public water utilities, expand service coverage and raise the quality of service, increase operating efficiency, provide alternative mechanisms of financing infrastructure investment, and reduce the burden on public budgets. This study describes and synthesizes the results of six representative experiences with private sector participation in providing these services in developing countries as well as economies in transition. Its principal objective is to draw lessons that can be applied to planning and implementing similar projects—both now and in the future. The cities and countries studied—Buenos Aires, Argentina; Cancún, Mexico; Cartagena, Colombia; Gdansk, Poland; Conakry and other urban areas in Guinea; and Santiago, Chile—have vastly different social, political, economic, and institutional conditions (Table 1). The six cases also illustrate a variety of different models of private sector participation, from service and managerial contracts to leases and concessions.

The reasons for soliciting private sector involvement varied across the cases studied:

- In Buenos Aires the Argentine federal government entered into a concession agreement with a private firm in 1993, as part of an extensive national privatization program undertaken to stabilize the economy.
- In Cancún the provincial and municipal governments were unable to meet the increasing need for water and sewerage services because of explosive growth in the tourist industry and the resulting population surge. The private sector was invited to meet the burgeoning demand under a concession contract.
- In Cartagena the municipal water and wastewater company had become identified in the public's mind with chronic inefficiency, political interference, and poor service. The national government had stepped in several times to restructure the company, but all efforts had failed. Ultimately, the mayor of the city decided to liquidate the company, and in 1995 a new company was constituted under joint public-private ownership.
- A mixed enterprise was also formed in Gdansk in 1992, in the context of democratic reforms and decentralization, to meet the need

Table 1. Principal Features of Private Sector Arrangements in the Six Cases

Case	*First year of private sector participation*	*Type of contract*	*Agency responsible for regulation*
Buenos Aires, Argentina	1993	Concession	Autonomous regulatory agency
Cancún, Mexico	1994	Concession	Regional agency
Cartagena, Colombia	1995	Operation & management	Municipal and national gov'ts
Gdansk, Poland	1992	Lease	Municipality
Conakry and 16 other towns, Guinea	1989	Lease	National agency
Santiago, Chile	1990	Service contracts	National agency

for system expansion as well as the need for better wastewater treatment facilities.

- In Guinea water supply was reaching less than 40 percent of the urban population in 1989. After failing to reform the public water company, the national government entered into a lease arrangement to provide water services for the capital city of Conakry and sixteen other towns.
- In Santiago public corporations were formed to operate water and sanitation services as autonomous commercial enterprises, with the state as the majority shareholder. Service contracts have been relied on extensively since 1979, and a comprehensive tariff system was developed to replace cross-subsidies with targeted subsidies funded by the central government.

Analysis of these six cases suggests that private sector participation in the water and wastewater sector is likely to result in sharply improved managerial practices and higher operating efficiency. However, it is unrealistic to expect the private sector in the short term to overcome all the inherited institutional and operational inefficiencies and to compensate for chronic underinvestment by the public sector. Moreover, the public sector's failure to establish clear regulatory frameworks and to implement adequate tariff regimes and subsidy mechanisms constitutes a palpable risk for the sustainability of private sector arrangements. Complementary reforms are required, especially in the areas of regulation, pricing of services, and financing.

Overall, private sector participation has led to improved service quality and expanded coverage. Management has been strengthened, productive efficiency has improved quickly, and sound commercial practices have increased revenues. In addition, cadasters have rapidly become more comprehensive, water losses have diminished, and attention to customers has improved significantly. Many of these initial successes have resulted from relatively simple management improvements that did not require large investments or sophisticated technologies. Private firms have shown a remarkable capacity to optimize the operation of existing infrastructure within a short time.

For example, in Buenos Aires, two and a half years after the private sector partner began operations, an additional 570,000 inhabitants have been connected to the water system (an increase of 9 percent), and 340,000 inhabitants have been connected to the sewerage system (a 7 percent increase). During the same period water production capacity was augmented from 3.4 million to 4.2 million cubic meters a day (an increase of 26 percent). Collection rates have improved in all six cases, by amounts ranging from 5 percent in Cancún to 60 percent in Cartagena. In Gdansk and Cartagena the private sector partner rapidly implemented new commercial and accounting systems. In Gdansk the private operator was able to raise the quality of drinking water to the standards of the European Union after just one year. In Guinea water supply coverage increased from 40 to 52 percent of the urban population between 1988 and 1994, and the share of urban households equipped with water meters increased from 5 to nearly 95 percent. The Santiago water utility has implemented a contracting policy that has promoted competition, reduced costs, and increased flexibility. More than thirty activities are currently under service contract, accounting for more than half of total operating costs.

Gains in operating and productive efficiency are not an automatic result of private sector participation, however. Unfavorable macroeconomic conditions, weak regulatory environments, and inadequate incentives can limit or nullify any gains in productivity. Moreover, initial gains and benefits, although important and positive, cannot by themselves compensate for the structural problems—ineffective public institutions, low productivity, low domestic saving rates, regressive tax systems, and extreme poverty and income disparities—that explain much of the poor overall economic performance in developing countries.

Consolidation of initial gains and the achievement of additional benefits will require effective institutional transformation and strengthened public administration. The success of reforms will depend on sustained, strong political commitment and on the support of supplementary reforms in three basic areas: more effective regulatory regimes; realistic and efficient tariff regimes, accompanied by direct subsidy mechanisms that increase the prospects for political and financial viability of tariff levels and make it possible to provide quality service to the poor; and development of innovative, tailored financial strategies and design of realistic investment programs.

Effective Regulation to Match Ends and Means

Effective regulation is the cornerstone of sustainable private sector participation. But the creation of a regulatory framework does not by itself guarantee effective regulation. It is fundamental to understand how institutions work, why they do not always adopt policies that outside observers recommend, and how this dynamic affects the performance of public utilities and private sector participants. Whatever the model of private participation adopted, the regulated companies and the regulatory authorities will face a steep learning curve as they seek to define their relationships with each other and work together to provide services equitably and efficiently.

Across the developing world there is an imbalance between the limited means and capabilities of public regulators and the capacity of experienced private operators. Strengthening public administration and regulatory institutions is crucial to reduce this imbalance and improve negotiating power.

Decentralization can tend to aggravate this imbalance. In countries that have shifted authority away from central government, many municipalities have assumed responsibility for providing public services. However, municipalities seeking to enter into equitable and sustainable arrangements with private partners may be seriously hampered by financial constraints, weak institutions, and low enforcement capacity. Frequently, the central government will need to provide financial support and regulatory guidelines to compensate for these weaknesses.

Further development of public-private arrangements may be impeded on the supply side by the limited number of experienced water companies. Competition in this industry is restricted to a few world players. Policies are needed to encourage the entry and development of local firms.

Initial efforts at private sector participation have been concentrated in major urban centers. But workable, long-run alternatives—both financial and institutional—are needed for small towns and cities as well.

Realistic Tariff Regimes

Most privatized utilities inherit inefficient tariff structures that give the wrong incentives to operators and consumers. If low tariffs cause private operators to lose money when providing service to the poor, access of the poor to services will lag. Incentives need to be corrected by making it profitable to serve the poor, and poor customers need help in paying for the cost of service through targeted subsidies explicitly included in government budgets.

Investment plans should be based on more realistic revenue projections. Greater attention should be paid to the effect of metering and full-cost pricing on consumption.

Evidence shows that political factors continue to influence tariff setting even after the private sector is brought in. The lack of clear policies and procedures for adjusting tariffs invites such influence. The methodology for calculating tariffs should be transparent and should specify objective criteria that trigger tariff adjustments.

Innovative, Tailored Financial Strategies

Under present conditions the risks of investment in the water supply and wastewater sector are significant, and the rewards uncertain. Given this imbalance, private firms are unlikely to be able to finance large investments in the sector. For the foreseeable future the bulk of financing will therefore come from cash generated by their own operations and from borrowings underpinned by a positive cash flow.

Private sector participation has already generated incremental resources for investment, although actual funding has lagged behind expectations. Wastewater treatment has proved particularly difficult to finance. Given the large benefits to society as a whole associated with pollution control, for which private investors are not fully compensated, public sector co-financing of these investments seems justified.

The model of private sector participation should be suited to the economy in question. One can differentiate two scenarios. In growing and stable economies, with reliable institutions and strong financial markets, private investment should be encouraged to take commercial risks and compete for credits in financial markets. This will foster efficient management and operations, expansion, rehabilitation, and system maintenance. In low-income countries with underdeveloped capital markets, low coverage, rapid population growth, and weak institutions, however, private companies will need to focus first on developing managerial and operating skills, increasing productive efficiency, developing comprehensive investment plans, and increasing creditworthiness. Here the private sector can contribute to reducing—but will not eliminate—the need for government financing. Finding the right mix between private and public sector investment will be the principal challenge.

The availability of long-term financing continues to be the critical constraint facing private investment in the water and wastewater sector. To a large extent that constraint is likely to be overcome only through financial sector reforms that encourage domestic saving and provide low-risk vehicles for infrastructure investment. The development of local capital markets, the expansion of investment guarantees, and the development of other innovative financial instruments to provide investors and private operators with access to credit remain important challenges for national governments and multilateral lending institutions.

The expansion of private sector participation will also be aided by reducing excessive transaction costs and the delays often incurred in designing and awarding contracts. International lending organizations such as the World Bank may contribute in several ways. They can prepare and disseminate guidelines for bid documents and model contracts that national and local governments can use as references. They can help governments set up regulatory and legislative regimes. They can provide timely financing of the services of independent, specialized consultants. They can develop training programs in specific areas of expertise for local officials, consultants, and domestic firms. Finally, they can design systematic monitoring mechanisms to assess current arrangements and the difficulties they encounter; the lessons drawn can help improve the quality of other initiatives, existing or envisaged.

The Sustainability of Models of Private Sector Participation

The reforms so far initiated and the private sector involvement already begun in the water supply and wastewater sector are a positive step. However, these changes will not necessarily be sustainable unless certain complementary requirements are met. To consolidate reforms, efforts must be accompanied by several conditions:

- *A sustained, favorable economic environment*. High inflation, falling real incomes, unemployment, and income inequality endanger sustainability.
- *Legitimate regulatory decisions and effective service provision*. The results of private sector participation must be perceived as beneficial by a large proportion of the population. Individual support and compliance with such reforms in the long term will depend critically on the delivery of promised services and benefits at reasonable cost.

- *Improved design of projects and targets.* Poorly designed private sector arrangements can have serious consequences for other projects by eroding credibility and deterring investors.
- *Complementary structural and institutional reforms.* Such reforms are necessary as well. Private sector participation in and of itself will not compensate for institutional and operational inefficiencies. Ineffective institutions endanger sustainability. Yet it is much more difficult to change institutions than to pass legislation, create new regulatory agencies, or design better tariff regimes. The probability of institutional change increases when the collective perception of effectiveness and fairness in the distribution of resources increases.

Needed domestic reforms include macroeconomic and structural reforms aimed at sustaining stable economic growth, combined with reforms to alleviate poverty and reduce inequality. Also crucial are reforms to the judiciary system that improve enforcement of regulatory frameworks. Reforms at the international level relate to the reorientation of multilateral organizations and better policy coordination between them and their clients. Without these reforms, the prospects for long-term sustainability of private sector participation in the water supply and wastewater sector in developing countries will remain in doubt.

1
The Context of Private Sector Participation in Developing Countries

The concept of private sector participation in public service provision embraces a wide range of policy approaches. At one end of the spectrum is outright privatization, or the "transfer of ownership and control from the public to the private sector, with particular reference to asset sales" (Hemming and Mansoor 1988). Other, less radical forms of private sector involvement include management and service contracts, lease arrangements, concessions, and build-operate-transfer arrangements (BOTs). The cases examined in this study represent a sampling of these various intermediate forms of private sector participation in the water and wastewater sector.

In developing countries, increased private sector participation in public services, including privatization of state-owned enterprises, is often part of a broader reform process whose goal is to resolve or preclude a macroeconomic or fiscal crisis. But there is a second, more immediate reason for the renewed interest in private provision: pervasive discontent with the performance of state-owned enterprises has reinforced the attractiveness of private sector participation as a way to reduce the economic inefficiencies associated with public management and ownership. In the wake of the 1980s debt crisis, private sector participation has gained support for yet a third reason, namely, that it offers a means of facilitating repayment of international debt, both through the proceeds expected from the sale of publicly owned enterprises and (as in the cases examined here) through reductions in government financial support for public services.

Strategies of private sector participation in public utilities have been promoted as a means toward achieving several goals: expanding service coverage and raising its quality, generating resources to finance future investments, increasing economic efficiency, reducing fiscal burdens, and introducing technological advances. Many governments have successfully leveraged the increased public discontent caused by decaying infrastructure systems, poor service delivery, and chronic inefficiency to win political acceptance of private sector involvement. Some governments have been able to incorporate strategies of private sector participation in water utilities into their overall economic stabilization programs.

The macroeconomic reforms carried out in a number of developing countries in the early 1990s produced some impressive results. In many countries inflation declined substantially, growth rates revived, foreign inflows of capital increased, and the fiscal environment improved. Yet despite these initial achievements, many reforming economies remain weak. In Latin America, for example, although the economic panorama varies widely—with growth rates in 1995 ranging from 8 percent in Chile to -7 percent in Mexico—growth on average is estimated to have declined from 4.6 percent in 1994 to 0.6 percent in 1995, the lowest rate since 1990. Meanwhile foreign investment in the region remains restrained, and social indicators have deteriorated. For quite different reasons, output in Eastern Europe and Central Asia has also fallen sharply, and high inflation and rising unemployment still place major limitations on the performance of these economies. The task of shedding seventy years of central planning and totalitarian government is taking longer and proving more daunting than originally believed.

Relying on simplistic and standardized prescriptions often leads to results substantially different from those predicted. The relative success of the initial macroeconomic reforms in many developing countries has actually disguised certain structural problems that remain a primary source of poor economic performance. These problems include ineffective public institutions, low productivity, increasing poverty and extreme income disparities, poor international competitiveness, low domestic saving rates, and regressive tax systems. Inadequate analysis of these problems may have helped create a climate of excessive optimism and unrealistic goal setting. Perhaps the main lesson from the reforms thus far undertaken in the developing world is the need to take a more penetrating look at the institutional, political, and socioeconomic realities within which reforms must be implemented.

This lesson is relevant to reform of the water and wastewater sector. The cases analyzed in this report, as well as similar experiences in other cities and countries, suggest that although the first experiments with private sector participation have yielded gains, including improvements in productive efficiency, they do not guarantee either that those gains will be sustainable or that the sizable resources needed to finance the necessary future investments will be forthcoming. In fact, technical and political constraints have undermined some of the reforms, and in many instances implementation has lagged behind stated intentions. This gap between expectations and results does not mean that significant progress has not been made. But it does suggest that some of the original targets set may have been unrealistic, and that the requisite financial resources—or the financial sector policies that could have generated those resources—may have been lacking.

Technical and managerial deficiencies in the water and wastewater sector of developing countries remain great. In the developing world as a whole, the percentage of the urban population connected to water supply systems increased only from 77 percent to 82 percent between 1980 and 1990; at the same time the absolute number of unserved urban dwellers rose by almost 15 percent (Munasinghe 1992). This means that in 1990 about 250 million urban residents in developing countries—a number equivalent to the population of the United States—lacked access to safe drinking water. In the absence of effective strategies, this number is likely to double by 2000. In all, about 1 billion people, urban and rural, in developing countries lacked access to safe water supply in 1990, and 1.5 billion lacked adequate sanitation facilities. In Latin America approximately 95 percent of domestic and industrial wastewaters are discharged without treatment. A grave threat to public health persists, as exemplified by the return of cholera to the region after an absence of more than 100 years. Similarly, the environmental damage caused by untreated wastewater discharged into natural recipients—rivers, lakes, and streams—is still rising.

Developing countries face a huge financial challenge: they have accumulated a massive backlog of needed investments not only in infrastructure but in other sensitive sectors of economic development as well. To meet only the demand in the water and wastewater sector, developing countries will need to invest between $30 billion and $60 billion per year over the next decade.

But the major challenge ahead for the developing countries in reforming their water and wastewater sector is to reconcile macroeconomic stability with more effective social policies linked to better provision of basic services. If output is falling, however, it is unlikely that these countries can muster the requisite investment in human capital and physical infrastructure, while at the same time stabilizing their economies through drastic reductions in public expenditure and heroic efforts to service their debts. Developing countries face the task of implementing a second wave of reforms that can consolidate their initial gains and address the problems posed by the endemic structural limitations of their economies.

2
Experiences with Private Sector Participation

This chapter describes the economic, legal, and institutional frameworks within which private sector arrangements were adopted in the six cases selected for study and summarizes the main features of each arrangement. The chapter ends with a brief description of several other relevant experiences with private sector participation in the water and wastewater sector, all of them in Latin America.

Buenos Aires, Argentina

Private sector participation in the water and wastewater sector in Argentina is part of an extensive privatization program intended to help stabilize the economy after the crisis of the 1980s. Service provided by the public company Obras Sanitarias de la Nación (OSN) in the capital, Buenos Aires, was deficient. OSN suffered from weaknesses typical of most public water utilities in the developing world: inefficient operation, weak commercial and financial management, poor maintenance, lax billing procedures, and excessive political interference. In May 1993 the federal government, with World Bank support and following an international bidding process, awarded a thirty-year concession contract to Aguas Argentinas, a consortium led by the French firm Lyonnaise des Eaux-Dumez (which would operate the water and wastewater system under the contract) and including Sociedad Comercial del Plata, Meller y Banco de Galicia (Argentina), Aguas de Barcelona (Spain), Compagnie Générale des Eaux (France), and Anglian Water (United Kingdom). Aguas Argentinas was established with capital of $120 million. To ensure the stability of the concession and proper incentives for good operation, the lead firm is obligated to retain at least 25 percent of the consortium's shares over the life of the contract. In November 1994 the International Finance Corporation (a member of the World Bank Group) became a partner in the consortium, with an equity share of 5 percent.[1]

The concessionaire is responsible for providing all water supply and sewerage services for the federal capital itself and fourteen districts of the province of Buenos Aires. Its responsibilities include commercial and technical operations, maintenance, and provision of all investment

Table 2.1. Performance Targets for the Buenos Aires Water and Wastewater Concession
(percent)

Year of concession	*Water coverage*	*Sewerage coverage*	*Sewage treatment*		*Unaccounted-for water*
			Primary	*Secondary*	
0	70	58	4	4	43
5	82	66	64	9	37
10	90	75	75	16	34
20	97	90	90	90	28
30	100	95	95	95	25

Source: Aguas Argentinas and Buenos Aires concession contract.

financing necessary to achieve certain performance targets designated in the contract (Table 2.1).

The contract does not specify how much investment will be needed. Instead, quality standards and performance targets are established for each of six consecutive five-year periods. It is projected, however, that meeting these targets will require an investment of some $4 billion. Personnel and fixed assets of the former public company were transferred to the concessionaire. At the end of the concession period all fixed assets are to be transferred back to the government in good working order.

A new regulatory agency, Ente Tripartito de Obras y Servicios Sanitarios (ETOSS), was created as an autonomous entity to control and maintain service quality, protect consumers, and approve and supervise the execution of expansion plans and investments according to contract specifications. A major responsibility of the regulatory agency is to review the tariff structure and oversee its compliance with the contract.

Cancún, Mexico

The story of Cancún is unique among the cases discussed here. In just twenty-five years Cancún has been transformed from a small fishing village into the most important tourist complex in Mexico, with more than 2.2 million visitors yearly and nearly a one-third share of the country's total tourist revenues. The city's population has grown seventeenfold, from 20,000 inhabitants in 1976 to 340,000 in 1995. This explosive growth led to serious deficiencies in the provision of public services and damaged the natural environment as well.

Following the constitutional reform of 1983, municipalities in Mexico became responsible for providing water supply and sanitation services. The municipality of Benito Juárez, where Cancún is located, delegated this responsibility to a state organization, the Water and Sanitation Commission (Comisión de Agua Potable y Alcantarillado, or CAPA). However, neither the municipality nor the state of Quintana Roo was able to meet the steadily increasing need for water and sewerage services in Cancún. In 1991, therefore, CAPA entered into a fourteen-year build-operate-transfer (BOT) contract with a private construction company, Desarrollos Hidráulicos de Cancún (DHC), a subsidiary of the conglomerate Grupo Mexicano de Desarrollo.

Under the contract DHC was to build and operate a new well production system to meet the increased water demand. However, the magnitude of the investments required for additional expansion and financing of wastewater treatment were well beyond the scope of the initial arrangement. In 1993, therefore, the state government, CAPA, and the municipal governments of Benito Juárez and neighboring Isla Mujeres invited four private firms to present proposals for the expansion, operation, and maintenance of the entire water and wastewater system in the cities of Cancún and Isla Mujeres, including wastewater treatment. DHC was awarded a thirty-year concession contract, which set five-year targets and efficiency standards for the expansion and improvement of services. The concessionaire created an operating firm, AGUAKAN, which took over operation of water and sewerage services on January 1, 1994.

AGUAKAN inherited all its staff and its tariff structure from the former operator, CAPA. Under the new arrangement the contract is regulated by CAPA, which supervises AGUAKAN's performance and approves tariff increases. AGUAKAN pays CAPA an annual concession fee of $1 million per year for the use of the existing infrastructure. It also pays the federal water agency an annual water extraction fee. AGUAKAN's profits come from total operating revenues after paying operating and maintenance costs.

The concession contract specified certain investments that AGUAKAN was to undertake. These investments, however, have not proceeded as stipulated, especially in the wake of the Mexican macroeconomic crisis of 1994 and 1995. As a result, the concessionaire has not met some of its targets, and this fact has eroded public support of Cancún's local government. In the hope of overcoming these difficulties—and of strengthening its regulation—the municipality of Cancún decided to become more closely involved. As a first step, in January 1996 the municipality created a technical board to oversee AGUAKAN's performance and participate in tariff decisions.

Cartagena, Colombia

Water coverage reaches an estimated 70 percent of Cartagena's population of 750,000, but 40 percent of the population served experiences rationing. Sewerage coverage is 58 percent. For many years the public municipal water and sewerage utility of Cartagena, Empresas Públicas Distritales (EPD), was notorious for chronic inefficiency, excessive political interference, and poor service delivery. The central government tried to implement a series of restructuring programs to improve its performance. All of them failed. In response to an unsustainable situation, in October 1993 the mayor of Cartagena decided to liquidate EPD and create a new, mixed-capital company to operate the system.

The local authorities, backed by the central government and with support from the World Bank, opened a bidding process in which three private operators participated. Only Aguas de Barcelona submitted a proposal deemed responsive to the community's needs. A new mixed-capital company, ACUACAR, was constituted in December 1994, with a share capital of $4 million. ACUACAR is owned jointly by the District of Cartagena (with 50 percent of its shares), Aguas de Barcelona (45 percent), and other private shareholders (5 percent). After six months of negotiations with the new mayor, ACUACAR and the city of Cartagena signed a twenty-six-year operation and management contract, and the company assumed control of the system in June 1995. Under the contract, the District of Cartagena continues to own the system's assets and is responsible for financing all future expansion of the network; ACUACAR itself undertakes no investment obligation. However, the district's capacity to provide the funds for service expansion is uncertain, and this could cause problems in the future. Aguas de Barcelona's agreed compensation consists of a fixed percentage of total revenues plus dividend distributions from ACUACAR's profits, as determined annually by its board.[2]

Gdansk, Poland

As a consequence of sweeping democratization, decentralization, and market reform in Poland, municipal investments began to increase rapidly in the early 1990s, following a major shift in responsibilities to the municipalities from the central government. In 1995 the population of the city of Gdansk on Poland's Baltic coast was nearly 500,000. The deficiencies of the city's water and sanitation system, coupled with the need for more efficient management to expand the system, eventually led to an arrangement with the private sector. Parallel to this initiative, the city needed to upgrade treatment of its wastewater to comply with the multinational effort to restore the Baltic Sea to its ecological equilibrium. In 1991 the municipal council of Gdansk ordered the liquidation of the regional water utility. Negotiations were initiated with a French company, Saur, for management and operation of the system. In July 1992 a mixed company, Saur Neptun Gdansk S.A. (SNG), was created. The city of Gdansk owns 49 percent and Saur 51 percent of the new company's equity of 7.75 million new zlotys (approximately $5 million).

A thirty-year lease contract was signed between the city of Gdansk and SNG in 1993. The city retains ownership of the relevant infrastructure and is responsible for capital investment and for financing, regulation, and setting of tariffs. The private operator, Saur, is responsible for operation and maintenance of the system, maintaining quality-of-service standards, and billing and collection. The arrangement thus basically follows the French *affermage* model. The city of Gdansk, through its municipal council, controls and regulates the performance of the company both as a shareholder in SNG and through the contract. This model implies a conflict of interest, however, and in fact the relationship between the city and SNG has been complex and tense. Economic conditions and regulatory procedures are not yet stable. A new municipal council, elected in mid-1994, criticized some aspects of the contract and decided to renegotiate it. In June 1995 the contract was amended to establish clearer procedures for and timing of annual tariff negotiations, sharing and control of information, and definition of a coherent annual operating plan consistent with politically viable tariff increases. A new remuneration formula for the private operator was defined on the basis of a fixed return on capital.

Guinea

Until the late 1980s Guinea had one of the least developed urban water supply sectors in West Africa. Less than 40 percent of the urban population had access to piped water. In 1995 the capital, Conakry, had a population of around 1.2 million, or 84 percent of the country's urban population. In August 1989, after attempts at restructuring the national water company had failed, the government entered into a lease arrangement for private sector operation of water services in the capital city and sixteen other towns.

Two new companies were created: a state-owned water authority, Société Nationale des Eaux de Guinée (SONEG), and a water management company, Société d'Exploitation des Eaux de Guineé (SEEG). SONEG owns the urban water supply facilities in the cities and towns covered by the lease. It is responsible for sector development, including planning and implementation of new investments, setting of tariffs, and servicing of the sector's debt. SEEG is jointly owned by the state (49 percent) and a private foreign consortium (51 percent) formed by the French companies Compagnie Générale des Eaux and Saur. The private partner was selected by the government after international competitive bidding on the basis of the minimum rate to be retained by the lease contractor on each cubic meter of water collected. SEEG holds a ten-year lease contract with SONEG, under which it is responsible for operating and maintaining urban water supply facilities as well as billing and collection. SEEG retains an agreed share of the tariffs it collects and pays the rest to SONEG as rent. The consortium also provides management services to SEEG through a separate management contract funded out of a fixed percentage of SEEG's revenues.[3]

Santiago, Chile

Most water and sanitation services in Chile are provided through public corporations established as autonomous commercial enterprises. The state is the major shareholder through its Corporación de Fomento de la Producción (CORFO). Santiago's public water and sewerage agency, Empresa Metropolitana de Obras Sanitarias (EMOS), was created in 1977 and transformed into a shareholder company in 1989. A system of concessions was instituted whereby the Superintendency of Sanitary Services grants public and private companies the right to provide water and sewerage services within the area under its jurisdiction. EMOS was awarded four different concessions, including for production and distribution of drinking water, sewage collection, and treatment of the collected sewage. EMOS' concession area comprises Greater Santiago and twenty-one periurban localities covering an area of 450 square kilometers and about 5 million inhabitants (approximately 40 percent of the Chilean population).

The most remarkable features of water and wastewater sector reform in Chile have been the extensive reliance on service contracts and the development of a comprehensive tariff system together with social policies to ensure affordable service to the low-income and rural populations. Since 1979 EMOS has outsourced to the private sector those activities where specialized private firms can realize economies of scale and technical improvements.

The profitability of Chilean utilities depends on their success in reducing costs below certain yardstick levels, which are set for every company that meets productivity and efficiency criteria. Tariff structure is based on efficiency criteria for a "model enterprise," taking into account long-run marginal cost. A variety of parameters are evaluated for each utility with reference to acceptable productivity and efficiency ranges. Water rates are reviewed every five years, after capital costs, service standards, investment plans, and other factors have been evaluated.

The government's social policies include positive actions to facilitate access to and use of water and sanitation services in periurban and poor settlements. As part of its water sector reform of the late 1980s, the Chilean government adopted a system of direct subsidies for low-income consumers. These subsidies are financed from the budget of the central government, and qualification is based on household socioeconomic criteria. Other social policies include loans and special payment plans for new connections and help for poorer families who apply for government subsidies.

Other Examples of Private Sector Participation

Other efforts to involve the private sector in water and wastewater services are under way in Latin America, from Mexico to Argentina. These experiences, although valuable, are not as well documented as the six cases that are the focus of this study. However, the difficulties that have emerged in some provide useful examples of the kinds of problems to which private sector participation is subject.

Private sector participation is spreading throughout Argentina. Apart from the arrangement in Buenos Aires, already discussed, projects have been undertaken in the provinces of Corrientes, Santa Fe, Formosa, and Tucumán. The first of these was in the province of Corrientes. In September 1991 a consortium led by the local companies Sideco Americana and Ingenieria Tauro, with a small equity participation by Thames Water (a U.K. firm), won a thirty-year concession for the provision of potable water as well as sewerage and sewage treatment in the capital (Corrientes) and nine major towns of the province. The concession covers a population estimated at 520,000. Since the consortium began operations, consumers have realized important benefits through improved service and expanded coverage (Table 2.2).

However, investment has lagged well behind the targets established in the contract. For example, investments in sewage treatment facilities estimated at $18 million have not been made. Estimated investment needs for the first fifteen years of the concession are about $75 million, but the concessionaire has faced difficulties in securing the necessary financing. Service costs are high because of low population density, but the tariff structure is rigid. Arrears have become a growing problem. Water tariffs have more than doubled, creating disincentives for bill

Table 2.2. Performance Indicators for Water and Wastewater Services in Corrientes Province, Argentina

Indicator	*Before private sector concession*	*September 1995*	*Percentage change*
Production capacity (millions of cubic meters per year)	58	54	-7
Water coverage (percentage of all households)	62	71	15
Sewerage coverage (percentage of all households)	30	37	23
Number of meters installed	1,800	86,000	4,700
Number of workers employed	583	223	-62

Source: Aguas de Corrientes, S.A. (November 1995).

payment. Billed output in 1995 was $85 million, of which only $64 million, or 75 percent, was collected.

In Tucumán province, Aguas del Aconquija, a consortium led by Compagnie Générale des Eaux with the participation of local companies, won a thirty-year concession in July 1995 to operate, maintain, and expand water and sewerage systems in the province. The capital, San Miguel de Tucumán, has a population of about 600,000, or 60 percent of the province's total. Water coverage in the province is about 73 percent, but sewerage reaches only 29 percent of the population. Some of the contract's targets are quite ambitious: water coverage is to be increased to 100 percent by year 6 of the contract, sewerage coverage to 100 percent by year 11, and primary wastewater treatment to 100 percent by year 6 as well. Financial requirements to achieve these goals have been estimated at $312 million, yet how this financing might be obtained is not yet clear. Tariffs have approximately doubled in less than one year, with the expected negative impact on bill collection.

A new provincial government under the control of the opposition party was elected three months after private operations commenced. In February 1996 Tucuman's provincial legislature was expected to accept a recommendation by government officials to rescind the concession contract. Negotiations between the provincial government and the private operator are taking place. Some major issues under discussion are the redefinition of the concession area, which would reduce the concessionaire's responsibilities in the main urban centers, and the feasibility of reducing tariffs in exchange for a corresponding reduction in planned investment.

Puerto Vallarta, a tourist center on the Pacific coast of Mexico, has about 150,000 inhabitants. A BOT contract for a wastewater treatment plant was signed between the municipal agency SEAPAL and the U.K. company Biwater in 1993. The private operator took over operations in 1995. The treatment plant was built to high design and operational standards. However, some operational, financial, and political difficulties have emerged. Operational problems are associated with fragmentation of responsibilities and difficulties in coordinating and balancing operations between the different components of the water and sewerage systems. Under the contract the municipality guaranteed a minimum sewage effluent of 600 liters per second, against a potential plant capacity of almost 1,000. This has not happened; instead the actual sewage effluent has been about 400 liters per second.

The system's financial difficulties are linked with political issues. The traditionally dominant party in Mexico, the Partido Revolucionario Institucional (PRI), was defeated in the 1994 municipal elections by the Partido de Acción Nacional (PAN). The new local government initially decided to avoid tariff increases. This policy weakened the financial capacity of the municipality. When the government tried to restore SEAPAL's financial viability by raising tariffs, the hotel sector sued, challenging the legality of the tariff adjustment process, and hotels refused to pay their bills directly to the municipality. Instead, payments were deposited in an account set up by the court. Its financial capacity thus eroded, SEAPAL has failed to meet some of the monthly payments due to the private company. To protect the contractor, the BOT arrangement had established a payment guarantee from Banobras, the Mexican public works bank, which is covering municipal liabilities temporarily. Negotiations to overcome these difficulties are under way.

BOT contracts are complex arrangements. Future private sector participation in Puerto Vallarta will demand a more refined conceptualization of projects, better coordination, and improved design parameters. Fragmentation of responsibilities for management, operation, maintenance, and investment among the different components of the water and sewerage systems (production, distribution, treatment) increases the risk of inefficiencies and lack of accountability.

In 1993 the municipality of Aguascalientes in central Mexico entered into a twenty-five-year concession arrangement with a consortium formed by the Mexican construction company ICA and Compagnie Générale des Eaux. The concession is for water and sewerage services to the city's population of 750,000. Mexico's 1994 currency crisis affected the concessionaire's financial capacity. Interest rates increased fourfold during 1995. In an attempt to restore the financial viability of the concession, tariffs were increased by nearly 60 percent. During the municipal political campaign of 1995, the opposition party, PAN, made renegotiation of the concession contract a major electoral issue. PAN won both the state and the municipal elections. A complex negotiation process is under way between the federal government, the municipal government, and the private operator.

Another important lesson from these experiences is that developing private sector initiatives in small towns and provinces with limited political power is quite different from doing so in the national capital or in large, politically influential cities. In fact, one factor that diminished the expected benefits of Tucumán's arrangement was its attempt to reproduce the Buenos Aires model. A key question is whether that model could be simplified to adapt it to different starting conditions in other countries or cities.

Attempts to introduce private sector participation in Caracas, Venezuela, and Lima, Peru, have so far been unsuccessful. In Venezuela the government announced in 1991 its intention to grant a twenty-five-year concession for water and sewerage service in Caracas. As in Buenos Aires, the concessionaire would have been responsible for any investments necessary to achieve the targets for service quality and coverage specified in the contract. An international competitive bidding process was undertaken, and five international consortia were prequalified. However, all five firms declined to bid, claiming lack of political support for the process as well as an absence of profit guarantees adequate to the risks. In Peru, after advancing decisively in project preparation and prequalification of bidders, the government postponed awarding the concession until 1997, thereby eroding the confidence of private firms in the political will of the government to engage in private sector arrangements.

Notes

1. For a detailed description of the stages of transition from public to private operation and of the concession arrangement, see Idelovitch and Ringskog (1995).

2. This component of the private party's compensation relates to a "management cost" fee. This fee is set at 2.85 percent of total revenues during the first year, increasing annually to 4.85 percent in year 5, and remaining constant thereafter to the end of the arrangement. Although this formula creates an incentive for effective rate collection, there is no incentive for cost reduction.

3. A detailed description of private sector participation in Guinea is provided by Triche (1990).

3
Level and Quality of Services

This chapter describes the effects of private sector participation on the level and quality of water and wastewater services in the six principal cases described in the previous chapter. The cases show that private sector participation did lead to improvement in the quantity and quality of service delivery. The direction of the observed changes is not surprising, for two reasons. First, as experience elsewhere has shown, in the short term, capital injections and sound technical advice lead to increases in coverage and improvement in water quality even in inefficient public water utilities. Second, the decision to contract with a private service provider is usually made in response to chronic degradation of existing systems and manifest incapacity of the public operator to overcome severe deficiencies: in Buenos Aires and Cartagena, for example, the public systems were under imminent risk of collapse. This low starting point could magnify the initial effects of any change. What is a welcome surprise is the speed with which private companies have been able to implement these changes. Even more important, however, is understanding how these changes can be consolidated and sustained.

Changes in the Six Cases

When Aguas Argentinas took over operations in Buenos Aires in May 1993, 6 million people had connections to piped water and 5 million to sewerage. By January 1996 the homes of an additional 570,000 of the city's inhabitants had been connected to the water system (a 9 percent increase), and those of an additional 340,000 inhabitants had been connected to the sewerage system (a 6.4 percent increase). Table 2.1 in the preceding chapter shows the targets established under the concession contract for Buenos Aires. It was expected that by the end of the first five-year period 1.6 million inhabitants would be added to the water network and 900,000 to the sewerage system. Compliance with all performance targets specified in the concession for the initial five-year period would entail an investment of around $1.2 billion, or $240 million per year.

Under public operation, the production capacity of the city's largest water treatment plant (San Martín) was 2.4 million cubic meters per day, with deficient quality. In 1995 plant capacity was increased to 3.0 million cubic meters per day, and works scheduled for completion by

Table 3.1. Performance Indicators for Water and Wastewater Services in Buenos Aires

Indicator	*Before private sector concession*	*December 1995*	*Percentage change*
Production capacity (millions of cubic meters per day)	3.4	4.3	27
Population served (millions)			
Water	6.0	6.5	8.8
Sewerage	4.9	5.3	6.4
Length of network (kilometers)			
Water	11,100	12,100	9.0
Sewerage	7,100	7,300	2.8
Water pipes rehabilitated (kilometers)	n.a.	500	n.a.
Sewerage pipes drained (kilometers)	n.a.	3,500	n.a.
Number of meters in service	30,000	170,000	460

n.a., not available.
Source: Aguas Argentinas.

mid-1996 will increase this figure by a further 0.3 million. For the first time in many years, no water rationing occurred during the last two summers (1994–95 and 1995–96) in Buenos Aires. Table 3.1 summarizes some of the major achievements of the Buenos Aires concession.

Aguas Argentinas has begun renovation of the existing wastewater treatment facility. The capacity of the plant remains the same, at 115,000 cubic meters per day, or about 4 percent of collected sewage. There is a considerable backlog in new investments for wastewater treatment, and discussions are under way over whether to revise the targets for this component of the system. In January 1996 the regulatory agency, ETOSS, contracted with an advisory group of international consultants, who produced in April 1996 a more detailed study of the technical and financial viability of the treatment investment plan.

Besides these achievements, the quality of potable water has been upgraded in Buenos Aires. A new water quality laboratory and equipment for quality control of drinking water and sewage effluents were commissioned. Water pressures have been optimized and physical leakage has diminished. Average repair time was reduced from two weeks before the new arrangement to forty-eight hours in 1995. Aguas Argentinas has also improved customer service: new payment systems were introduced allowing customers to pay bills at nearly 900 payment counters, including the company's own customer service centers, commercial banks, and financial institutions.

Cancún suffers from an imbalance in public service provision between the city's residential and tourist areas. In the residential zone, water and sewerage coverage are 63 percent and 35 percent, respectively, and all residents there endure severe water rationing (ten hours per day on average). However, the tourist zone enjoys 100 percent coverage in both water and sewerage, and service is reliable and continuous. Wastewater treatment coverage is likewise 100 percent for the tourist area, versus 26 percent for the residential areas. This remains unchanged even with the entry of the private concessionaire. The hotel sector represents 0.5 percent of all water users yet consumes about 40 percent of water produced and accounts for about 70 percent of AGUAKAN's operating revenues.

The concession contract set as its goal to reduce this imbalance by increasing water coverage for the whole of the city to 95 percent by the end of 1996 and sewerage coverage to 95 percent by the end of 1998. Although AGUAKAN has taken some steps to improve service delivery and coverage, there is an investment backlog with respect to the targets established in the concession contract. AGUAKAN is currently contracting to add 10,000 new water connections to the system, which will increase water coverage to 74 percent by the end of 1996. The company also added 79 kilometers of pipeline to the sewerage network, to increase coverage to 45 percent. These works, however, remain inactive because of lack of funds to build pumping stations and wastewater treatment plants.

Water production increased from 44 million to 47 million cubic meters per year in 1995, but unaccounted-for water has remained high at 51 percent of output. The contract set as its target to reduce unaccounted-for water to 30 percent by the end of 1996—an elusive goal given the lack of universal metering. As a first step, AGUAKAN installed thirteen water production meters and some 16,000 domestic meters, augmenting the share of metered customers from 54 percent in 1994 to 86 percent in 1996.

In Cartagena the public utility had made no investments during the eleven years prior to turning operations over to the private sector. As a result, service delivery is still poor. When ACUACAR took over operations in June 1995, new commercial and accounting systems were implemented within a few weeks. During the first three months of operation, ACUACAR computerized all administrative workplaces and opened seven new customer service centers. One consequence has been a drop in billing complaints from 4 percent at the start of operations to 2 percent by September 1995, three-fourths of which were resolved immediately, compared with 62 percent under public provision. Maintenance and rehabilitation programs were implemented for water and sewerage networks and channels as well. Quality of water has improved through better use of chlorination.

The investment challenge is great: estimates are that about $250 million will be needed over the next five years. ACUACAR is currently undertaking an $8 million emergency investment program, scheduled for completion before the end of 1996. This program is targeted to benefit 34,000 households. An investment priority will be to reduce unaccounted-for water from 52 percent to 40 percent in the first year and to 35 percent by the end of second year.

In Gdansk the private operator, SNG, succeeded in optimizing the performance of existing treatment plants and raising the quality of drinking water to the standards of the European Union by the end of its first year, exceeding the contract's provisions. SNG has also constructed a modern laboratory for effective control of drinking and wastewater effluents and has implemented a related program for control of industrial wastewater. More than 300 industries (nearly 100 percent of the city's total) have signed contractual arrangements with SNG, effectively adopting standard regulations for industrial effluents.

Although the population served and the number of water and sewerage connections in Gdansk did not change significantly, average water production fell more than 20 percent between 1993 and 1995. This reflects a sharp reduction in average demand, which dropped from 277 to 212 liters per day in the same period as a consequence of tariff increases and meter installation. Within a short time, SNG prepared a comprehensive ten-year investment, rehabilitation, and maintenance program for the city's water and sewerage systems. Although investments remain below expectations, the total length of water and sewerage pipe systems has increased 20 percent and 3 percent, respectively, over the last three years. The number of pipe breaks fell from 763 in 1992 to 550 in 1995 (to 56 occurrences per 100 kilometers per year), and average repair time was reduced to ten hours—substantially less than the twenty-four hours established in the contract.

In Guinea major gains in service availability and quality were achieved during the first five years of the lease contract. In part these gains accrued from substantial investment in new water supply, which increased production capacity from 7.5 million to 28.7 million cubic meters per year between 1988 and 1994.[1] This increase in supply capacity, combined with progress in rehabilitation and maintenance, brought about a substantial increase in the share of the population with access to safe water, from 40 percent in 1989 to 52 percent in 1994. Water connections increased from 16,500 in 1989 to 33,500 in 1995. Metering has increased from 5 percent to nearly 95 percent of all connections. Less encouragingly, as many as 10,000 connections (about one-third) are now inactive as a result of nonpayment. Rates increased from $0.24

Table 3.2. Performance Indicators for the Santiago Water and Wastewater Utility (EMOS)

Indicator	*1990*	*1991*	*1992*	*1993*	*1994*
No. of connections (thousands)					
Water	837	867	905	944	985
Sewerage	777	807	868	915	956
Coverage (percentage of all households)					
Water	99	100	100	100	100
Sewerage	91	93	95	97	97
Water production (millions of cubic meters per year)	462	453	466	469	475
Unaccounted-for water (percent)	28	27	27	24	22
Pipe breaks per hundred kilometers per year	39	39	38	35	31

Source: EMOS.

to $0.90 per cubic meter between 1989 and 1995.

EMOS, the public water and sewerage utility in Santiago, is today the best-performing water utility in Chile and quite possibly the best in Latin America. Operational and financial indicators reveal solid performance: water and sewerage coverages are 100 and 97 percent, respectively; the ratio of employees to water connections is low (2.1 per 1,000); metering is near 100 percent; and unaccounted-for water is below 22 percent. Net profits in 1994 were 37 percent of total sales, or 11 percent of total assets. Yearly dividends averaged $21 million during the last three years. Table 3.2 shows recent trends in some of these indicators.

Perhaps the major challenge EMOS faces over the next two decades will be financing and implementing an ambitious program for wastewater treatment in Santiago. Investments estimated at around $600 million will be needed. As part of this plan, in 1993 EMOS put in operation a pilot wastewater treatment plant and in 1994 a wastewater collector, which receives 60 percent of the city's sewage.[2] Financing and cost recovery are now being studied.

Conclusions

Some conclusions and general trends can be drawn from these six cases:

- Private operators have been able, to a greater or lesser degree, to both expand the quantity and improve the quality of water and wastewater services. The greatest improvements are observed in

Buenos Aires and Santiago, whereas the Cancún concession shows little progress. Public sector participation in Guinea showed impressive results initially, as a result of significant capital injection from an International Development Association credit, but further gains have been elusive. These differences seem to endorse the view that the magnitude of these gains may depend not on the particular model of private sector participation but on the quality of the incentives perceived by the water companies themselves. (The role of incentives in utility performance is discussed in Chapter 5.)

- Many of the initial improvements were achieved by introducing relatively simple management and operating procedures that do not require large investments or sophisticated technologies. Private firms have shown a remarkable capacity to optimize the operation of existing infrastructure shortly after taking control.
- Private sector participants have given first priority to increasing the flexibility and improving the reliability of water and wastewater systems and to equipment rehabilitation, inspection and mapping of distribution systems, regulation of network pressures, identification of required rehabilitation and other activities to reduce water losses, and implementation of systematic control procedures.
- On the commercial side of the business, immediate improvements include new billing and collection systems, updating of cadasters, rapid incorporation of users into the commercial system, systematization and decentralization of information, and better consumer service.

Notes

1. This investment, totaling some $58 million, was raised with the participation of the International Development Association, the African Development Bank, the French Caisse Centrale de Coopération Economique, and the European Investment Bank.

2. The capacity of this treatment plant is 0.2 cubic meters per second, which represents less than 2 percent of the total sewage collected in Santiago (15 cubic meters per second).

4
Cost of Services to Consumers

The design and implementation of efficient, equitable, and sustainable tariff systems remain a high priority and a great challenge for public and privately operated water utilities in developing countries. The cases reviewed in this study indicate that, although progress has been made, rigidities continue to impede the imposition of adequate tariff systems.

Tariff structures under private sector arrangements have followed no uniform trend. Different levels and structures are observed, depending on the inherited tariff structure, investment needs and financing sources, the type of contractual arrangement, the regulatory framework, and socioeconomic and political factors. Nonetheless, some common directions are beginning to emerge, and these are identified below. This chapter discusses the main challenges and constraints to implementing more efficient tariff regimes in the water utilities of developing countries.

Tariff Trends

Tariff increases following the entry of private sector participants are likely to be substantial, especially during the transition from public to private sector provision. There are several reasons why this is so. First, water and wastewater systems in developing countries are characterized by low rates of coverage and large and increasing investment needs for expansion and rehabilitation. Second, before the private sector participant is brought in, poor financial practices usually cause water rates to fail to cover operating costs. Third, under the new arrangements, private capital costs must be included in the economic cost of services.

With private sector participation, necessary tariff increases can partly be offset by gains in productivity. However, it is unrealistic to assume that productivity gains alone will be enough to generate the large increases in funding that are needed. In the case of large and inefficient utilities, private sector participation can lead to significant cost reductions, which in turn increase the likelihood that water rates can be reduced to reasonable levels for consumers (see Chapter 5). Buenos Aires provides a good illustration: despite an increase of 13 percent in the second year of the concession, current water and sewerage rates are still 17 percent below what they were before the private sector entered the picture (Box 4.1).

This will not always be the case. Elsewhere in Argentina, in the provinces of Corrientes and Tucumán, tariffs roughly doubled after private sector participation was introduced. In Chile the resulting nominal tariff increases between 1990 and 1994 were significant compared with the old rates: after applying marginal cost principles, the average increase across Chile's thirteen water supply regions was 76 percent, with a maximum increase of 463 percent and a minimum of 7 percent. During the same period the real tariff increase in Santiago was 41 percent. These increases were gradually implemented over a four-year period, from 1991 to 1994. By late 1994 average tariffs for water and sewerage services ranged from $0.29 per cubic meter in Santiago to $1.02 per cubic meter in Region II in the northern part of the country (Figure 4.1).

In Guinea the average tariff rose from $0.24 per cubic meter at the beginning of the lease period to $0.90 in 1994—an increase of 270 percent (Brook-Cowen 1996). The new tariff level is high by both regional and international standards. In the past, however, low water rates did not cover operating costs and depreciation. As part of the lease arrangement, the government agreed to raise tariffs gradually to a level that covered operating expenses (including depreciation) and yielded an acceptable financial return on system assets and on the lease contractor's equity.

Box 4.1. The Buenos Aires Tariff Regime

In Buenos Aires the private sector concessionaire, Aguas Argentinas, adopted the tariff structure inherited from the former public utility, OSN. This regime is not based on actual consumption, is regressive, and provides no incentive for efficient use of water. Two main categories of service are defined: nonmetered and metered. Rates for nonmetered services are based primarily on the area of the property, its location, and the quality of construction of its buildings.

For metered services rates are calculated on the basis of two components: a fixed fee of 50 percent of the fixed rate defined for nonmetered consumers, and a charge of $0.55 per cubic meter for water consumption in excess of 30 cubic meters over a two-month period.

The winning concessionaire was selected on the basis of the lowest water rate bid. The winning consortium offered a rate 27 percent lower than the existing rate at the time of bidding. Under contract provisions these rates are to be reassessed every five years, based on the next five-year investment plan.

Two tariff modifications have occurred in the context of the move to private sector provision. The first, a rate increase of 8 percent, was implemented just prior to the call for bids. A second increase of 13.6 percent was approved in late 1994. Its purpose was to finance additional investments not included in the original five-year plan.

Figure 4.1. Targeted and Actual Tariffs for Water and Sewerage in Chile

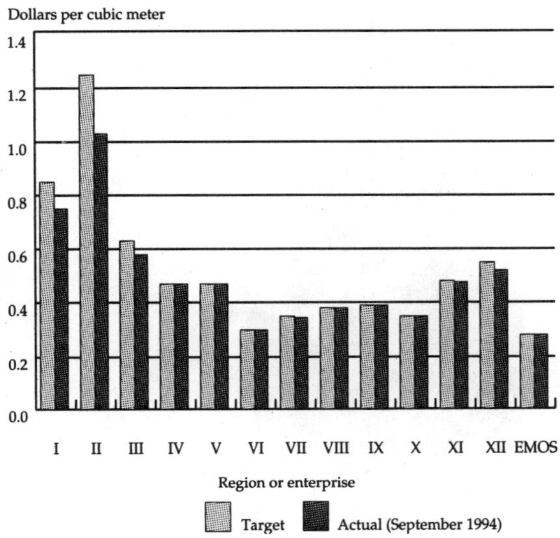

Source: Superintendencia de Servicios Sanitarios, Chile.

In the United Kingdom (England and Wales) and Murcia, Spain, the evolution of water rates under private sector participation proved interesting. A study of the tariff behavior of thirty-nine private water and sanitation companies in the United Kingdom (Rees 1995) shows that water rates are expected to increase between 20 and 120 percent in real terms over ten years after privatization (Figure 4.2). It should be emphasized, however, that water utilities in the United Kingdom must meet the water quality standards of the European Union, without subsidies.

The situation is different in Murcia, where water rates were reduced in real terms by 4 percent between 1989 and 1992, after a private sector arrangement went into effect. The decrease was not a direct result of private sector participation but was instead a consequence of public subsidies.[1] However, the private company's good managerial practices were a cardinal factor in obtaining financial support to develop an ambitious investment program. The company's cumulative investment in the period from 1989 to 1995 was about $48 million, of which $36 million (76 percent) came from external sources. The investment program for 1995 and 1996 included $50 million in financial aid from the European Union. Aguas de Murcia's after-tax profit was $2 million in 1993.

Figure 4.2. Expected Tariff Increases over Ten Years among Water Companies in the United Kingdom

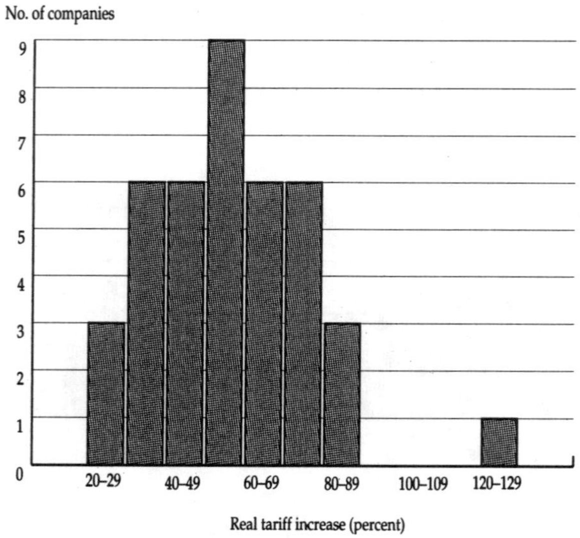

Note: Estimates are for the first ten years after beginning of private sector participation.
Source: Rees (1995).

Barriers to Implementing Efficient Tariff Systems

An efficient tariff system requires that prices reflect the value of the resources and services produced. An efficient system must also be functional, in the sense that it is understandable and perceived as fair, politically feasible, and capable of influencing the behavior of consumers in a predictable way. In only a few cases have local governments or private companies adopted the policies necessary to achieve efficiency in tariff regimes in their water and wastewater sectors. Chile is perhaps the only case in which a coherent set of legal and institutional transformations accompanied reform (Palominos 1994).

Failure to Link Tariff Regimes to Productivity

In many of the cases reviewed here, tariff regimes are not linked to productivity gains or investment requirements. An interesting issue is how to create appropriate tariff regimes that incorporate profitability and efficiency incentives for private operators and encourage the optimal use of resources by consumers. Once a system has been designed with the right incentives, the issue is how to determine the degree to which the operator has achieved an optimal and feasible cost structure, as well as how these costs are incorporated in the level of tariffs. Finally, if productivity gains are achieved, decisions will have to be made about how to distribute the benefits. In the United Kingdom and Chile, for example, where private sector participation in public utilities has proved profitable, there is evidence that the balance favored shareholders and executives at the expense of consumers (Rees 1995).

In Gdansk, after the 1994 municipal election the new city council decided to renegotiate the operation and maintenance contract. A new compensation mechanism for the private operator was established on the basis of a fixed percentage of capital. This formula is less likely to lead to operational efficiency gains to the extent that the operator's compensation is not linked to profits.

Low Metering Levels

Low metering coverage is another important limitation. A basic principle of economic efficiency is that consumers should pay on the basis of their actual consumption, which implies metering. Table 4.1 shows trends in metering coverage in some of the cases reviewed. The challenge is to increase metering coverage to give consumers the opportunity to save money by reducing consumption, while at the same time avoiding giving the operator a disincentive to install meters—a concern both in Argentina and in the United Kingdom.

Lack of metering is a problem in Buenos Aires. About 90 percent of residential consumers are not metered. These customers have no incentive to use water efficiently. Average consumption in Buenos Aires is estimated to range from 500 to 600 liters per day, compared with 260 liters per day in Santiago and 220 in Bogotá, Colombia. Under the concession contract, however, Aguas Argentinas has little incentive to

Table 4.1 Metering Coverage before and after Private Sector Participation in Selected Cases
(percentage of all connections)

City	*Before*	*After (December 1995)*
Buenos Aires	2	8
Cancún	54	74
Cartagena	47	62
Conakry and 16 other cities in Guinea	5	95

Source: Author's calculations.

promote efficient water use by residential consumers—at least in the initial stages of the concession, when pressures for capacity expansion are low. A massive increase in metering coverage could have a negative effect on net revenues, given the difficulty of reducing operating and investment costs in the same proportion as the expected reduction in consumption. The key issue is how to design a revenue-neutral tariff, at least for the transition period. The concessionaire and the regulator are looking into the problem. Meanwhile they have agreed to install a minimum of 80,000 meters (7 percent of Aguas Argentinas' water customers) per year.

Distorted Tariff Structures

In Cancún as in Buenos Aires, the concessionaire, AGUAKAN, inherited the existing tariff system from the state operator. This tariff structure is based on cross-subsidies between the hotel, industrial, and residential sectors. Such a regime introduces strong distortions. Hotel sector tariffs, for example, are subsidized at consumption levels between 70 and 1,500 cubic meters per month but become excessively high for consumption above 15,000 cubic meters per month. As a consequence, some hotels are studying the possibility of disconnecting from the system and installing desalinization plants. Similarly, tariffs for the industrial sector are heavily subsidized for the first 200 cubic meters per month. It is estimated that 85 percent of industrial consumers receive subsidies on all their consumption. By the end of 1995 the average water tariff for the hotel sector was $1.50 per cubic meter, while the average water price for residential consumers was about $0.25 per cubic meter.

This regime reflects a clear conflict of objectives: on the one hand, there is a contractual commitment to provide services to the poor; on the other, a distorted tariff structure introduces disincentives for private investment in low-income areas and generates resistance to paying bills in the hotel sector. Tariff adjustments above inflation must be submitted for approval by the regulatory agency, CAPA. However, the contract explicitly prohibits real tariff increases for the hotel sector during the initial five-year period, except in the case of an exchange rate crisis. Water rates in Cancún have been adjusted for inflation over the last five years.

Like its counterparts in Buenos Aires and Cancún, Cartagena's ACUACAR inherited its tariff regime from the former public utility, EPD. As part of the transition strategy, ACUACAR is committed to keeping this tariff regime in place during the first two years of operations. Colombia has a rather complex tariff structure defined at the national level and comprised of a fixed charge and three different strata of increasing per-unit charges for consumption. Residential consumers are divided into six socioeconomic categories. A 50 percent sewerage surcharge is added to the water bill. Prices are structured to provide cross-subsidies across income groups. However, the distribution of water and sewerage subsidies is not progressive.[2] Table 4.2 presents the tariff structure and water prices for Cartagena as of October 1995. Colombia's National Regulatory Commission has recently developed a new methodology for setting tariffs based on efficiency criteria, which went into effect in June 1996.

An interesting problem that emerges here is the difficulty of applying marginal cost principles or increasing utility revenues under distorted tariff structures. As can be seen in Table 4.2, about 79 percent of residential households in Cartagena belong to the low- and medium-low income categories (strata 1-4), and only 5 percent are in the highest income category. The average tariff is $0.40 per cubic meter in the residential sector. This means that at least 72 percent of connections and 50 percent of consumption are at prices below the average tariff. However, providing services for the remaining 30 percent of the unserved population—the poorest—would lower the average tariff even more, creating disincentives for the private sector provider to serve the poor. To overcome this situation, customary practice advocates additional tariff increases for higher-income groups and the industrial sector. However, price elasticities in these sectors are significantly higher than in the low-income categories, because industries have greater scope for substituting alternative supplies; thus the attempt to expand the level of subsidies might have adverse effects on a utility's finances.[3]

Table 4.2. Water Consumption and Tariffs Paid by Type of User and Household Income Category in Cartegena, October 1995

User category	*Share of all connections (percent)*	*Share of consumption (percent)*	*Average tariff (dollars per cubic meter)*
Residential stratum			
1 (poorest)	12	8	0.12
2	23	14	0.18
3	37	28	0.29
4	7	5	0.45
5	9	8	0.58
6	5	6	0.86
Commercial and industrial	6	21	0.60
Official	1	10	0.34
Total	100	100	0.40

Source: ACUACAR and author's calculations.

Lack of Clear Mechanisms for Setting Tariffs

The lack of systematic mechanisms for setting tariffs also limits the implementation of more efficient tariff systems. In Cartagena, Cancún, and Gdansk, contracts for private sector participation do not establish systematic tariff-setting mechanisms. In Gdansk, for example, since the conversion to private sector provision took place in 1992, four annual water tariff increases have occurred (Figure 4.3).

The municipal council and the private operator renegotiate a new rate every year through a complex and time-consuming process. Tariff adjustments have been subject to political considerations. Successive tariff increases below the rate of inflation (which has averaged 30 percent in Poland over the past three years) have undermined the financial capacity of the private lessee, SNG. By late 1995 SNG and the city council were again involved in a complex tariff negotiation. Ultimately the council approved a tariff increase of 20 percent, setting an average water and sewerage price of $0.80 per cubic meter for residential consumption. The private operator is expecting approval of a second increase by mid-1996, but as of March 1996 the council had not made a decision. These uncertainties have slowed proposed investments.

Figure 4.3. Water and Sewerage Tariff Increases in Gdansk Compared with Inflation

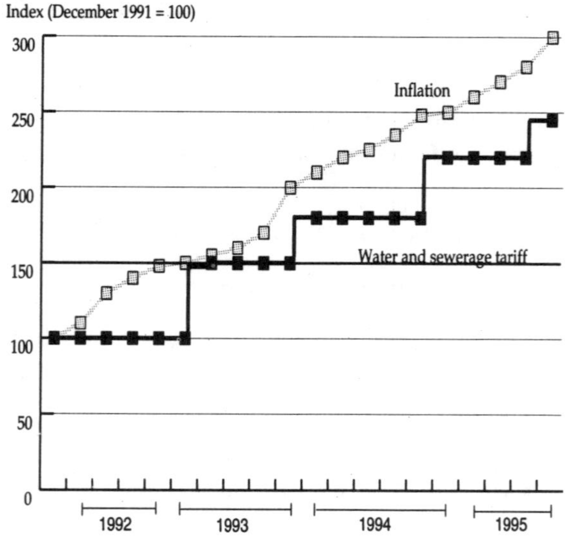

Note: For 1995 one U.S. dollar was worth 2.4 new zlotys.
Source: SNG.

The Need for Consistency

Investment programs need to be designed using sensible pricing criteria. In some cases of private sector participation a tendency has been observed to overestimate the expected revenues from tariff increases that theoretically will serve to finance ambitious investment programs, without a realistic assessment of the demand response. The introduction of tariff increases combined with expanded metering may dramatically alter consumption patterns. These variations in consumption can have different outcomes: on the one hand, they might reduce the utility's revenues; on the other hand, more efficient use of resources diminishes the pressure for investment to expand the network. However, when most consumers are unmetered, as is usually the case before private sector participation, estimates of the elasticity of consumption demand are unreliable. In countries facing high inflationary pressures, in which output and real incomes are falling, or in which a large proportion of consumers are poor, substantial tariff increases may dramatically affect the financial viability of investment programs.

Gdansk is an interesting example in this regard. Economic transformation in Poland has not been without unfavorable consequences: since 1989 real wages and real pensions have declined nearly 15 percent (World Bank 1995b, p. 4), unemployment has risen, and inflation has eroded savings. Relative prices of basic commodities, such as housing, food, and heating, which were previously subsidized, have increased substantially. Although water rates have not been raised in real terms, annual increases in nominal tariffs have had what SNG's management describes as an "unforeseen" impact on water consumption patterns. During 1992–95 total water consumption decreased by 33 percent in the city. In some districts the impact was even more dramatic. For example, in the district of Morena, where price adjustments were combined with an increase in metering coverage from 5 to 85 percent, water consumption fell by 52 percent in the same period (Figure 4.4). A direct consequence of this declining trend in consumption was a decrease in SNG's expected revenues, which in turn weakened its financial capacity and that of the municipality as well.

An important lesson that emerges from these experiences is the need to reconcile in a more realistic way the potential revenues from tariffs with investment plans. Various studies (for example, Cestti, Yepes, and Dianderas 1996) indicate that the income elasticity of water consumption can range between 0.1 and 0.5. These elasticities are higher when the water bill represents a large share of total monthly household income. Price elasticities are on the order of -0.3 to -0.6 for residential consumers. These variations impose a limit on the revenues that public or private

Figure 4.4. Water Production and Metering Coverage in Gdansk's Morena District

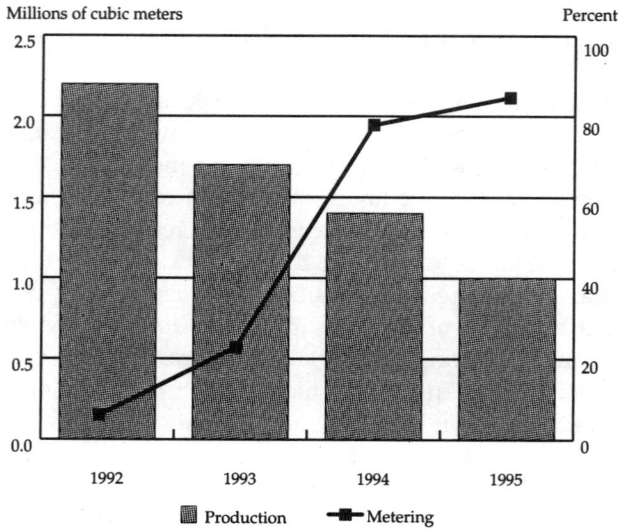

Source: SNG.

firms can raise from tariff increases. These limits vary according to the circumstances of each country or city, but they need to be reflected in the design of realistic investment plans. The case of Tucumán illustrates this situation: just six months after the concession contract was signed, Compagnie Générale des Eaux (the lead firm in the private sector consortium) and the provincial government were in the process of renegotiating the contract. One alternative under consideration was to reduce tariffs in exchange for a corresponding reduction in the investment plan.

The Importance of Subsidies and Additional Sources of Funds

When additional funds for investment are required beyond the limits imposed on revenues by structural constraints, the needed financial resources must come from sources other than tariffs (for example, government credits, subsidies, and cofinancing schemes). In such cases a coherent, direct subsidy structure (like that of Chile; Box 4.2) is a fundamental tool to ensure the political and financial viability of tariff regimes that accurately reflect the economic value of the resources and services produced. This is especially so when investment programs or expansion of coverage need to be accelerated, when projects include social objectives, and in the case of wastewater treatment projects. The design of innovative financial programs and incentives to promote the required investments remains a major challenge for local and national governments, multilateral lending organizations, and the financial sector in general.

Social and economic indicators in Guinea are among the lowest in the world. GNP per capita is around $450 a year. As noted above, after the advent of private sector participation the average customer tariff rose more than sixfold, to $0.90 per cubic meter in 1994. As a consequence, almost 10,000 connections (one-third of the total) are now

Box 4.2. Chile's Water Subsidy System

In Chile a system of direct subsidies managed by the central government has contributed to the political and financial viability of the tariff system. Each year the central government allocates a budgeted amount for subsidies, which are to be distributed according to preestablished social and economic criteria among the country's regions and municipalities. Municipalities verify the socioeconomic condition of families requesting the subsidy and classify them accordingly, with priority for subsidies given to the poorest.

The subsidy applies to fixed and variable charges for water and sewerage services for the first 20 cubic meters of monthly consumption and ranges from 25 percent to 85 percent of the water and sewerage bill. Once the central government transfers the funds to the municipalities, the municipalities use these funds to pay the water utility directly. Subsidies may be renewed for consecutive three-year periods. The total value of these subsidies, which cover about 450,000 families (18 percent of all Chilean consumers), is about $25 million.

In the case of EMOS, the water utility for the metropolitan Santiago area, subsidies total about $4 million per year, an amount equivalent to 2.3 percent of total billing. In addition to the subsidies, EMOS provides loans to help poorer families pay for water connections. A water connection costs between $200 and $800; a typical loan for such a connection would require a 15 percent down payment, with monthly payments over five years at commercial interest rates.

This form of subsidy is fully compatible with all forms of private sector participation, and its use might facilitate the implementation and sustainability of such projects. This is not to say that transferring this or any model is an easy process. The Chilean subsidy scheme relies to a large extent on strong local administrative capacity coupled with high governmental ethics—two assets that are not always easily transferable.

shut off because of nonpayment. At that price, the cost to a household of 20 cubic meters of water per month—a reasonable estimate of consumption for a developing country—is around $18. By comparison, at the same level of consumption, households in some representative European cities paid in 1992 (for water alone) $4 (Milan), $5 (Rome), $15 (Paris), and $17 (London) (Yepes and Dianderas 1994, p. 20). There is no easy explanation for Guinea's high rate. What is clear, however, is that the current level of tariff has led to a higher default rate—and stronger incentives for illegal connections. Commercial losses are in fact increasing. Accounts receivable stood at 200 days of revenue in 1995 (Brook-Cowen 1996). It is important to emphasize that during the initial years of the lease contract, the implementation of tariff increases was eased through a subsidy mechanism: the shortfall between tariff revenues and operating cost, as well as debt service, was covered through a direct subsidy funded by an International Development Agency credit on a declining basis, as tariffs increased.

Lessons

Several lessons can be drawn from these experiences with water pricing:

- Regardless of the model of private sector participation used—and given the existence of highly underpriced services, plus the large capital investment required for expansion and rehabilitation of water and wastewater systems in developing countries—there have been strong upward pressures on tariffs.
- Despite a consensus on the need to promote more efficient tariff structures that give the right signals to investors and consumers alike, strong rigidities slow their application. Many private companies have inherited or adopted inefficient tariff structures.
- In many cases there has been a tendency to set unrealistic goals, especially during the initial stages of private sector involvement. Investment plans should be more realistic and consistent with the potential for revenue generation from tariffs. Greater attention should be given to assessing the impact of metering and pricing on consumption patterns.
- Highly distorted, cross-subsidized structures continue to provide disincentives to expand services to the poor. Incentives have to be created for private companies to serve the poor, and these customers must be helped to pay for services through targeted subsidies explicitly included in government budgets.
- Tariff setting following private sector participation continues to be

heavily influenced by political factors. Although most private sector arrangements anticipate some mechanisms for tariff adjustment, these are not always transparent. The lack of clear policies and procedures for tariff adjustment invites external interference. An important lesson is that tariff changes should be transparent, well defined, and, if possible, self-adjusting on the basis of easily explained principles. Guidelines should specify how often tariffs will be adjusted, the process for adjusting them, and the methodology or principles to be used.

Notes

1. Aguas de Murcia receives financial support from the European Union, the Spanish Ministry of Public Works, and the Comunidad Autónoma de Murcia, and it benefits from investments made by the Mancomunidad de los Canales de Taibilla and the Confederación Hidrográfica del Segura. See Hervas (1996).

2. A World Bank study (World Bank 1994a) indicates that at least 70 percent of the connections required for the system as a whole to attain the level of coverage presently enjoyed by the highest income groups (about 1 million water and 1.8 million sewerage connections) would have to go to the poorest 40 percent of Colombian households. Moreover, in Colombia's main cities, households in the lowest income decile spend up to 9 percent of their income on water and sewerage, whereas the general population spends only 1 to 2 percent.

3. A recent study of Canadian industries found water price elasticities for the manufacturing sector ranging from -0.5 to -1.2 (Cestti, Yepes, and Dianderas 1996).

5
Gains in Performance and Efficiency

This chapter examines the efficiency gains that have been achieved in the six cases of private sector participation under review. Gains in economic efficiency comprise increases in *productive efficiency* and improvements in *allocative efficiency*. Increased productive efficiency means producing more output with the same or fewer inputs—in other words, at lower cost. Such gains do not necessarily correspond to gains in allocative efficiency, or improvements in the allocation of resources. Allocative efficiency requires that prices and tariffs more closely reflect scarcity values, and it requires a functional tariff structure. As was noted in Chapter 4, there is little evidence that such changes occurred under private sector participation in these cases, at least in the early stages. Although this deficiency is not inherent in private sector participation, what is clear is that gains in allocative efficiency do not happen automatically in such settings. They are more likely to occur if complementary reforms are introduced to ensure a stronger correlation between the cost of the service provided and the price charged to consumers.

Table 5.1 presents some basic operating and financial indicators for five of the six cases. In most cases, private sector participation led to significant gains in productive efficiency. However, where macroeconomic conditions were unfavorable (as in Mexico and Poland) or where the private firm lacked appropriate incentives (as in Cancún), these gains were small or absent. Broadly speaking, productive efficiency was raised by improving the quality of managerial decision making, reducing political interference in daily operations, and making managers responsible to shareholders. With private sector participation, commercial systems improved rapidly and collection rates rose. As a consequence, cash from operations increased significantly, and with it the firm's creditworthiness. However, improvements in collection rates and cost reductions tend to be large during the initial stages of new private sector arrangements, and these benefits diminish over time. Further gains in productivity—and the savings that result—become increasingly marginal. In similar fashion, all the companies in the sample improved their planning and control systems, analytical accounting systems, and contracting and procurement procedures. Service standards and access to technological developments also improved, and consumers now appear to be better informed about services and costs.

This chapter examines three major aspects of the performance of private companies in the water and wastewater sector: the capacity of these companies to increase revenues and reduce costs, the role of appropriate incentives, and labor force management.

Revenue Increases and Cost Rationalization

In Buenos Aires before establishment of the private sector concession, the annual revenues of OSN, the public water utility, barely covered operating costs, with nothing left for maintenance or for new investment. The billing and collection system was inadequate, and many customers were unregistered. After the private concessionaire, Aguas Argentinas, began operations, the combined effect of revenue gains from expansion, an improved commercial system, and a tariff increase of 13.6 percent caused revenues to jump by 76 percent, from $38 million a month in 1993 to $67 million in 1995. Complementary measures included incorporating about 30,000 illegal connections into the system and updating and reclassifying users. In addition, Aguas Argentinas has been able to reduce energy costs by 10 percent and the cost of chemicals by 44 percent. Revenue increases together with cost reduction strategies allowed the consortium to turn a loss of $26 million in 1993 into a profit of $26 million in 1994, which doubled to $54 million in 1995.

In Cartagena during the first six months of private operation, ACUACAR increased revenues and reduced costs. The collection rate improved from 50 percent to 82 percent, while metering increased from 56 percent to 61 percent. Gross operating revenues (on a half-year basis) increased by 25 percent. Personnel costs fell by about 70 percent after the implementation of an early (voluntary) retirement program financed by the central government, which reduced the number of workers from 1,200 under the former public utility to about 400.

The public water utility in Gdansk had enjoyed neither administrative autonomy nor control over its financial resources. Its revenues were usually diverted to other sectors without a clear rationale. The private lessee, SNG, implemented a new commercial system that led to significant improvements in collection rates. However, tariff increases fell short of inflation, causing a real decline in operating revenues of 15 percent between 1993 and 1995. This decrease was offset by a reduction in operating costs by 15 percent during the first eighteen months and by an additional 5 percent during 1993-95. The municipality has benefited from these savings. The proportion of total revenues transferred from SNG to the city increased from 18 percent in 1992 to 32 percent in 1995. In 1995 SNG transferred about $5 million to the city. Creating effective

Table 5.1. Operating and Financial Indicators for Five Cities before

Indicator	*Buenos Aires* Before PSP	*Buenos Aires* 1995	*Cancún* Before PSP	*Cancún* 1995
Operating revenues (millions of dollars)	230	385	25	21
Gross income[b] (millions of dollars)	-8	133	15	11
Working ratio[c]	1.05	0.65	0.4	0.47
Collection rate (percent)	80	6	85	88
Unaccounted-for water (percentage of total)	n.a.	n.a.	51	51
No. of employees	7,450	4,250	518	526
Employees per thousand water connections	6.4	3.3	12.8	11.4

Note: PSP, private sector participation; n.a., not available.
a. Data are for the first six months of operations (July-December 1995).
b. Total operating revenues minus net operating costs (excluding depreciation and interest payments).
c. Net operating costs divided by operating revenues.
Source: Author's calculations.

incentives for further cost reductions has been difficult, however. The private operator has expressed concern that all savings were being "appropriated" by the municipality.

As in the case of Cartagena, the scheme adopted in Gdansk represents a great improvement in administrative and operative terms over the previous situation. Management improved significantly in both cases; the municipalities control the savings generated from operations and decide planning and investment programs. Both cities can now anticipate their budgets for coming years based on tariff negotiations. Although these gains are not enough to cover all financing needs, they have made a positive contribution.

The Role of Appropriate Incentives

The case of Santiago's public water enterprise, EMOS, supports the argument that, with appropriate incentives, publicly owned companies can be efficient. EMOS' operational and financial indicators indicate that this is a well-run utility, with robust performance indicators. Table 5.2 shows trends in some of these indicators during 1990-94. As a public enterprise, EMOS was able to take the necessary actions to fulfill legal and regulatory

and after Private Sector Participation

Cartagena		*Gdansk*		*Santiago*	
Before PSP	*1995*[a]	*Before PSP*	*1995*	*1990*	*1994*
10	12	24	19	60	130
-2.5	0.9	7	5	29	71
1.2	0.89	0.71	0.74	0.52	0.45
50	82	80	98	90	94
52	50	n.a.	20	28	22
1,200	385	1,030	800	1,717	1,945
14	4.5	10.3	8.0	2.1	1.9

requirements pertaining to quality and continuity of service. Among the key factors that account for EMOS' superior performance are its outsourcing of services to private firms, the continuity of its management, strong leadership and accountability, and the presence of a comprehensive and efficient tariff structure, backed by a system of direct subsidies to low-

Table 5.2. Operating and Financial Indicators for the Santiago Water and Wastewater Utility
(millions of dollars except as noted otherwise)

Indicator	*1990*	*1991*	*1992*	*1993*	*1994*
After-tax profits	15	19	29	36	53
Real investment	19	34	51	44	46
Dividends paid	5	17	18	20	26
Operating costs per cubic meter of output[a]	0.18	0.20	0.21	0.23	0.22
Unaccounted-for water (percentage of total)	28.1	26.9	26.8	23.7	21.9

a. Includes depreciation.
Source: EMOS.

income consumers that helps ensure its effective implementation.

EMOS' contracting policy has promoted competition, reduced costs, and increased flexibility. EMOS does not contract out those activities considered "strategic" (for example, billing and collection, and accounting) or any other service that its own staff can perform at a lower cost. Contracts are normally awarded through competitive bidding for a period of two years and can be extended at EMOS' convenience. More than thirty activities (valued at $13 million a year) are currently under service contract. These activities account for about 52 percent of total operating costs, excluding depreciation. EMOS reduced its ratio of personnel costs to gross operating revenues from 27 percent in 1993 to 25 percent in 1994, in part by reducing the number of employees per 1,000 water connections from 2.1 to 1.9 (a 5 percent cut). The number of employees per thousand water supply *and* sewerage connections was slightly below 1.0 in 1994.

Two factors help explain EMOS' increased productivity. First, the profit-sharing plan it has offered to its unionized workers is attractive. EMOS has three unions. The latest collective bargaining agreement, negotiated in 1995, established that unionized workers would receive 16 percent of any increase in profits over those of the previous year. The second factor is a matter of survival. The military regime during its final years in power made clear its intention to privatize most Chilean water utilities. Chile's new democratic government, elected in 1990, continued the privatization process. Emerging local conglomerates and some government agencies expressed great interest in privatizing sectors such as mining, ports, and water and sanitation services. This system of indirect competition or permanent "threat" encouraged publicly owned water companies to improve their performance and productivity in an effort to neutralize the arguments for privatization. An interesting challenge is how to maintain these incentives over the long term. The government is now considering the partial divestiture of some of its thirteen water companies by means of asset sales rather than concessions.

In contrast, after two years of operations Cancún's private water operator has not achieved significant productivity gains. Some improvements have been made, but much work remains to be done to maximize management and operating efficiencies. Part of this underperformance can be traced to the 1994–95 currency and financial crisis in Mexico, which sent inflation spiraling upward from 7 percent in 1994 to 52 percent in 1995.

Mexico's recent crisis is only part of the story, however. Three other factors have compromised AGUAKAN's performance. The first is the weakness of the regulatory scheme. CAPA, the former public operator now charged with regulating AGUAKAN, has no previous experience as a regulator, and AGUAKAN's contract does not provide the appropriate package of incentives for the operator to reduce many of the inefficiencies inherited from the public utility. The second factor is a "cost-plus" syndrome whereby high costs and some inefficiencies are simply passed on to consumers through higher tariffs, ensuring acceptable returns to the operator. Thus, the expected benefits of reducing inefficiencies did not compensate for the costs incurred. The third factor is the concessionaire's lack of knowledge about the water and wastewater sector. Lacking experience in managing water and sewerage systems, AGUAKAN's parent company (a construction firm) had no clear understanding of the key indicators to be improved and monitored. At least in the initial stages, the concessionaire devoted more attention to the potential for construction activities than to the key administrative and operating issues that characterize the water sector.

An important lesson is that private sector arrangements will achieve operational efficiency only if the private operators are given the right set of incentives. This implies not only reasonable rewards but the acceptance of risk on the part of the private contractor. In a protected environment, with a distorted set of incentives, private firms may be less than efficient and may not perform any better than public agencies.

Leases as well as operation and management contracts require effective coordination and a clear allocation of commercial risk between the government, as investor, and the private party, as operator. In Guinea, for instance, the incentives established in the lease contract adversely affected the relationship between the government water authority, SONEG, and the water management company, SEEG. The latter is responsible for new connections and operation and maintenance of those portions of the pipeline network that do not exceed 160 millimeters in diameter. SONEG is responsible for financing and contracting for extensions and major rehabilitative works over the entire network. Thus, both entities have some capacity to influence both the rate at which connections are added and that of reductions in unaccounted-for water. The practical result was that the water supply system did not improve or expand as quickly as expected, new connections to the system were added slowly, and unaccounted-for water remained high. At the beginning of the arrangement, water losses were estimated at 40 percent; these grew to 62 percent in 1993 before declining to 47 percent in 1995 (Brook-Cowen 1996).

Each of the two entities has a tendency to attribute slow progress to the other's failures: the government agency blames the relatively slow pace of new connections on the private operator's reluctance to make connections to existing extensions to the network, while the operator asserts that much of the demand for new connections is located in areas where the government agency has yet to invest in network extensions. "The net effect has been that the lease has come to approximate a management contract, under which the commercial risk borne by the private party is reduced, and its incentives for efficiency improvements are correspondingly weakened" (Brook-Cowen 1996).

Management of Human Resources and Labor Force Reductions

Private sector initiatives can have a powerful effect—symbolically, at least—on the balance of political and economic power between the public and the private sector. In many countries experienced labor unions and other worker organizations have accumulated a significant share of political power through decades of negotiations. Careless management of this component of the social contract can reduce the expected benefits of private sector arrangements, particularly in countries with ethnic, religious, or intense social tensions. For example, insufficient attention has been given to developing coherent strategies for the implementation of effective and equitable retirement plans for public employees. Labor opposition to private sector involvement has been usually managed by one of three basic strategies: offering generous retirement programs including substantial incentives for workers to take early retirement; keeping a large proportion of the work force employed in the new private company; or reserving a certain proportion of equity shares for former employees.

In the six cases reviewed here, new retirement plans were implemented only in Buenos Aires and Cartagena, where operating and institutional problems had reached unsustainable levels before the private sector partner came on the scene. In Buenos Aires, Aguas Argentinas' work force was reduced by nearly 50 percent within six months after the start of operations. As part of this process, 7,450 employees were transferred from OSN to the concessionaire. The central government funded a voluntary early retirement program, which about 1,600 employees accepted at a cost to the government of about $32 million. Another 2,000 employees were separated through a similar voluntary retirement plan financed by the concessionaire at a cost of $50 million. Later the concessionaire and the union negotiated a new collective bargaining agreement and a 40 percent increase in the average wage. Under contract provisions, the concessionaire reserved 10 percent of Aquas Argentinas' shares to be bought by its employees. Between 1993 and 1995 the operating company provided its staff with about 150,000 hours of training, benefiting more than 12,000 participants.

(Many employees took part in more than one session.)

In the case of Cancún, AGUAKAN was forced under the provisions of the contract to hire all staff from the former public operating company, CAPA. After two years of private operations the number of employees has remained relatively constant. The current high ratio of eleven employees per 1,000 water connections suggests a problem of excess personnel. Again, there are reasons to believe that weak regulation and a distorted system of incentives might have induced the private operator to emulate the public operator's performance. Personnel costs, however, are not excessive compared with other water companies in the region. In 1995 these costs represented about 32 percent of operating costs, which is below the ratio for Aguas Argentinas (43 percent) or for Santiago's EMOS (57 percent).

Prior to negotiations with the private sector in Cartagena, about 600 employees accepted a voluntary retirement program funded by the national government and the municipality. The remaining staff continued to decrease through attrition, to approximately 400 employees. These would be hired by ACUACAR under its agreement with the municipality. Before the private operator came on the scene, pension obligations were covered by the public utility, EPD, out of its operating revenues. Thereafter these costs, which represent about five months of ACUACAR's operating revenues (or approximately $8 million per year), were assumed by the municipality. Together with ACUACAR's gains in productivity, this shift of pension liabilities to the municipality plays a major role in explaining the company's positive financial results. During its first six months of operations, ACUACAR earned net profits of about $1 million, compared with a deficit of about $5 million reported by EPD before ACUACAR took over operations.

Gdansk is one of the few Polish cities whose water utility has experienced labor retrenchment. The private operator has developed a gradual but comprehensive program to promote the creation of microenterprises among its employees. Those who participate in voluntary retirement programs are encouraged to form small private companies, which are then contracted by SNG to perform specific tasks such as cleaning and maintenance. This is a noteworthy achievement given the legacy of socialist legislation and the sensitivity of unemployment in Poland.[1] In addition, SNG has trained more than 1,500 participants in over seventy different areas of expertise during 1993–95. SNG's promotion of private microenterprises among the former employees of the public utility offers an interesting option that other cities considering private sector participation would do well to explore.

Lessons

Three main lessons can be drawn from this discussion:

- In all six cases there was a notable improvement in commercial operations after the private operator came on the scene. In several cases—those of Buenos Aires, Cartagena, and Santiago—private operators were able to increase operating revenues and reduce operating expenses, thus boosting profitability. However, such gains are not an automatic result of private sector participation. When macroeconomic conditions are unfavorable or when the private firm lacks appropriate incentives, greater profitability or gains in productivity are small or nonexistent. In Cancún and Gdansk, operating revenues and gross incomes have fallen in real terms since the private contractor began operations. Institutional, political, and macroeconomic constraints have contributed to these unfavorable results.
- Private sector participation does not automatically guarantee internal efficiency. Operational efficiency will come about only if private operators are given the right set of incentives. This implies not only reasonable rewards but the assumption of risk by the private contractor. In a protected environment, with a distorted set of incentives, private firms may not be efficient and could well perform poorly.
- Private operators have promoted reductions in staff, alleviating the financial burden of the utilities. Although this is a sensitive issue with deep political and social implications, little attention has been given to the design of coherent strategies and practical guidelines for implementing equitable and predictable negotiation processes in the context of private sector participation.

Note

1. Unemployment has increased dramatically in Poland since its transformation into a market economy: "In December 1989 the main unemployment office in Warsaw paid benefits to five people. A year later, twelve months into their "big-bang" reform, more than a million Poles were unemployed, and by mid-1993, 3 million" (Barr 1994, p. 1). Official unemployment in Poland was about 15 percent in 1993.

6
Financial Aspects

Financing investment in the water and wastewater sector remains a challenge: to meet demand, developing countries will need to invest around \$60 billion per year, or \$1.2 billion every week, during the next ten years. This will mean increasing water supply and sanitation investments from less than 0.4 percent to about 1.0 percent of their combined GDP in the next decade.

The cases reviewed in this study indicate that, given the current balance of risks, incentives, and rewards, the private sector is unlikely to invest its own resources in the water and wastewater sector in developing countries. Most private operators have made only minor investments in operational improvements, choosing those with obvious high returns. In cases involving concessions, where private operators have overall responsibility for services, including capital investments, the major source of investment has been cash flow generation and borrowings supported by that cash flow. Private sector participation clearly has produced gains in productive efficiency and generated financial surpluses. These gains are positive and real. But they will not be sufficient to fund all the sector's investment needs.

The water and wastewater sector offers fewer competitive options than do other infrastructure sectors. Assets required per dollar of annual revenue are greater in this sector than in telecommunications, transport, or electric power,[1] and these assets amortize over long periods and have limited or no resale value. Moreover, the financial performance of water and sewerage utilities in developing countries is often worse than that of other infrastructure sectors. A recent World Bank study (World Bank 1994b) found that cost recovery in the sector is only around 35 percent. Part of the problem is the ambivalent nature of water as both an economic *and* a social good, which sends a conflicting message. In addition, both the weakness of domestic capital markets and increasing evidence that public policy concerning water utilities remains high on the political agenda even after the private sector enters the picture have led private investors to become more selective in their participation in projects in this sector.

Domestic and foreign credit operations pose special challenges in developing countries. In practice, lending operations are restricted under current models of private sector participation. First, because neither concessionaires nor operating companies own the assets they are charged with managing, they cannot use those assets as security. This is not an insurmountable obstacle, although it requires finding ways to use the revenues generated from those assets as collateral instead. This mechanism works well if the tariff-setting and adjustment process is predictable and if revenue-pledging arrangements are legally in place. Second, the private sector participant's equity is usually small relative to investment needs, and in many cases its use is explicitly curtailed.[2] Access to long-term loans requires sophisticated capital markets, financially responsible companies, and well-structured projects. The private sector can play a valuable role when these are present. But where capital markets are incipient, as they are in most developing countries, the transition from public financing to long-term private financing will take time and ingenuity (Serageldin 1994).

The use of loan guarantees to support private project lenders has met with difficulties in its practical application. World Bank guarantees, for instance, require that countries provide counterguarantees to the Bank. However, many developing countries have decentralized their public sectors, transferring responsibility for the water and wastewater sector to the local level. This has made central governments reluctant to guarantee municipal lending operations unless a reliable revenue-pledging arrangement can be put in place, which is more the exception than the rule.

The magnitude of the financial challenge and the specific constraints of the water and wastewater sector require the design of coherent, long-term financial strategies. The private sector can and should play a decisive role in these strategies. But its role is a limited one and should be managed in a more pragmatic and balanced way, according to the specific challenges, risks, and opportunities.

Buenos Aires and Cancún

The investment targets set in the case of the Buenos Aires concession are the most impressive of the six cases studied. As previously noted, the contract contemplates an investment program of approximately $4 billion over the thirty-year concession period, including an estimated $1.2 billion (30 percent of the total), or about $240 million per year, in the first five years.

Figure 6.1 shows the distribution of projected investment in Buenos Aires over the life of the concession. These figures serve to underscore the tendency of many private sector projects to concentrate most of the investment program in the initial stages. This is also the case for the concession at Cancún and in the Argentine provinces of Tucumán and Corrientes.

To date, two and a half years after being awarded the concession, Aguas Argentinas has invested about $320 million of the $1.2 billion

Figure 6.1. Projected Investments of the Water and Sewerage Concessionaire in Buenos Aires

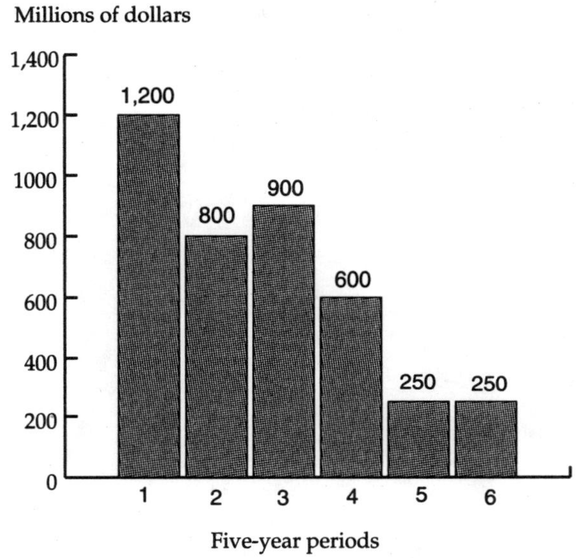

Note: The first year of the first five-year period covered from May 1993 to May 1994.
Source: Aguas Argentinas.

projected for the first five-year period. The concessionaire's initial investments were financed mainly by its initial share capital ($120 million), internal cash generation, the first loan of $172.5 million from the International Finance Corporation (IFC), and short-term debt.

The financial plan for the first five-year period contemplates obtaining financing from long- and short-term loans (62 percent of the total), cash flow from operations (28 percent), and equity (10 percent; Table 6.1). Aguas Argentinas has completed negotiations with the Inter-American Development Bank (IDB) whereby the Argentine government will on-lend the proceeds of a twenty-year loan of $98 million from the IDB originally intended for the state-owned public utility, OSN. This IDB loan, expected to be made available during 1996, carries a government guarantee. The financial plan also includes the second IFC loan of $213 million, approved in January 1996. In addition, the concessionaire is engaged in discussions with the European Investment Bank for a long-term loan of approximately $90 million for construction of the North wastewater treatment plant. IFC participation and the government guarantee of the IDB loan have played a decisive role in enhancing the concessionaire's creditworthiness.

Although investments for expansion of water supply and sewerage coverage in Buenos Aires have been satisfactory, there is a considerable

Table 6.1. Sources of Funds and Planned and Actual Investments of Private Sector Participants in Buenos Aires and Cancún

Participant	*Millions of dollars*	*Percentage of total*
Aguas Argentinas (Buenos Aires)		
Sources of funds		
Equity	120	10
Cash generation	330	28
Loans		
IDB	98	8
IFC I (1994)	172	14
IFC II (1996)	213	17
EIB (proposed)	90	8
Other	177	15
Planned investments, 1993–98	1,200	100
Actual investments, 1993–95	320	27
AGUAKAN (Cancún)		
Sources of funds		
Concessionaire	12	10
Cash generation	10	8
Planned investments, 1994–2000	120	100
Actual investments, 1994–95	22	18

Note: IDB, Inter-American Development Bank; IFC, International Finance Corporation; EIB, European Investment Bank.
Source: Aguas Argentinas, AGUAKAN, and author's calculations.

backlog of investments for wastewater treatment. One of the more important discussions regarding the investment program is Aguas Argentinas's proposal to study alternative ways of reaching the targets for sewage treatment established in the contract. The concessionaire has an incentive to defer these investments: for example, it has been estimated that its savings in capital costs from delaying construction of the Berazategui wastewater treatment plant are about $100,000 per day. The regulatory authority will be guided in its decision by the final recommendation of the recently contracted advisory group of international consultants.

The viability of Aguas Argentinas's investment program will depend in large measure on the continued availability of long-term financing at reasonable rates. Financial projections indicate that, by 1998, the concessionaire's total long-term debt will reach about $750 million. To fulfill its financial obligations and meet the targets established in the contract, Aguas Argentinas will have to maintain steady progress in increasing revenues from 1996 onward. Essential to meeting this objective will be the achievement of additional productivity gains, continuity in implementing expansion programs, and the results of the tariff negotiation planned for the second five-year period.

Foreign investors in private sector arrangements continue to perceive devaluation of the local currency as a substantial risk. An alternative strategy to attenuate this risk would be to promote lending operations in local currency, but this option is not always feasible, especially in countries where capital markets are weak. Maturities and interest rates are seldom compatible with the protracted construction periods characteristic of the water and wastewater sector. One concern of the Buenos Aires concession is the considerable risk associated with the sustainability of the exchange rate parity system in Argentina. Although the concession contract explicitly provides for extraordinary tariff revision in the event of a major devaluation, Aguas Argentinas has repeatedly requested that the tariff be denominated in dollars rather than in Argentine pesos.

Other complex issues are the need to provide service to the low-income population and the growing demand to accelerate expansion of the system to areas where the concession had scheduled connection for later stages of the contract. About 35 percent of the unconnected population of Greater Buenos Aires are below the poverty line. Explicit lump-sum subsidies to support these low-income consumers may have to be considered. Bringing future investments forward would necessarily affect tariff levels. Technical and financial alternatives are being explored. One of these is to encourage Aguas Argentinas to concentrate on installing the large primary network and allow local governments to raise funds and manage the installation of secondary distribution systems, under previously defined standards.

The financial challenge faced by AGUAKAN in Cancún is also formidable. Estimates of needed capital expenditures for the period 1994–2000 are about $120 million. AGUAKAN's pace of investment has been below expectations, however. During the first year of the concession (1994), the company invested about $15 million, part of which was funded by its private sector parent, DHC. However, in 1995, after the Mexican peso devaluation, the private operator stopped funding investments, which fell by 53 percent to only $7 million in 1995, for a total of $22 million in 1994 and 1995 (Table 6.1). In 1996 AGUAKAN initiated preliminary discussions with the IFC, seeking its participation as a shareholder in the Cancún concession.

The service and coverage targets established in the concession for the first five-year period appear quite ambitious. Table 6.2 shows current and targeted coverage rates. Given current limitations in time and financial resources, it is unlikely that the concessionaire will accomplish these goals.

Table 6.2. Targeted and Actual Water and Wastewater Coverage in Cancún
(percent)

Service	*Target for 1996*	*Actual, December 1995*
Water	95	62
Sewerage	95[a]	37
Wastewater treatment		
Hotels	100	100
Residences	None[b]	26

a. Target is for 1998.
b. Contract does not specify a target.
Source: Aguakan.

Cartagena and Gdansk

Under the private sector participation agreements in Cartagena and Gdansk, the municipality is responsible for new investment, which is to be undertaken with public funds. Given the fiscal stringency facing many local and national governments in developing countries, some have contended that this type of arrangement lacks the necessary mechanisms to ensure the financial resources needed for expansion and rehabilitation. However, it is also true that, given the current balance of incentives and risks, the private sector would be prone to concentrate its investment on those large urban centers and medium-size cities with more attractive economic prospects. But it could be argued that, in the specific context of the Cartagena and Gdansk contracts, it was unrealistic to expect the private sector to assume risks for operation and management as well as for future investment. One issue that needs further analysis is the duration of operation and management contracts in cases where the private sector has no responsibility for investment. The terms fixed in the cases of Gdansk (thirty years) and Cartagena (twenty-six years) seem too long for this type of model.

Investment needs are large in both Cartagena and Gdansk, especially for wastewater treatment and disposal. Estimates indicate that Cartagena will need to invest $250 million in the next five years and Gdansk approximately $100 million in the same period (including investment in wastewater treatment). Under current arrangements there is no guarantee that the municipalities will come up with the financial resources needed.

In Cartagena, during its first six months of operation ACUACAR invested about $4 million as a part of a $8 million emergency investment plan. Funds came from cash flow generation, equity, and commercial credits. Although the contract calls for creation of an investment fund to finance system expansion, this instrument is unregulated, and its mechanisms for financing and functioning are not clear. Present tariff levels are inconsistent with investment needs. The private operator accepted the current low tariffs for the initial two years as a way to facilitate negotiations. More realistic tariff levels are expected by 1997, when the two-year period ends.

To procure the required resources, the city of Cartagena sought financial support from the Colombian government through grants for the sanitation component. In addition, the private operator and the city have entered into preliminary negotiations with the World Bank for a loan of about $140 million to ensure continuation of the ongoing water supply and sewerage project.[3] Meanwhile ACUACAR is working toward the definition of a comprehensive master plan study for the water supply and sewerage systems. The plan is expected to be completed by the end of 1996. The municipality faces a critical financial situation—in fact, it is essentially bankrupt. Thus, the financial viability of the investment plan will depend on ACUACAR's access to long-term financing as well as on the resources available to the Colombian Co-financing Fund.

In Gdansk, as in Cartagena and Guinea, the municipality assumed financial responsibility for investment. In 1993 the private operator, SNG, developed a comprehensive ten-year master plan for investment, rehabilitation, and maintenance of the water and sewerage systems. However, financial restrictions affected the pace of investment. Under the private sector arrangement, funds for capital investment came from operating surpluses. During its three years of operation, SNG's annual investments have averaged about $4 million per year.

Perhaps the major investment challenge for Gdansk in the coming years is the construction of its wastewater treatment plant, which will permit the improvement of water quality in the tourist zone on the Baltic Sea. The estimated cost of this project is around $50 million; however, the financial plan is not yet clear. Mobilization of debt finance has been difficult. In 1993 a proposed loan from the European Bank for Reconstruction and Development was canceled at the government's request. Constraints associated with additional tariff increases and the weakness of Polish capital markets have further reduced the prospects for successful loan negotiations.

Savings from improved operation and management in Cartagena and rent payments in Gdansk have contributed positively, although not sufficiently, to generating the necessary resources for investment and to improving creditworthiness. What is clear is that appropriate financing strategies require municipalities to set and secure tariff levels that will enable recovery of capital costs and contribute adequately to investment. As this has proven difficult, it appears that the only way to finance investment needs is through some combination of public and private finance.

Guinea

The experience of Guinea provides an example of the challenges and opportunities faced by a poor country in which the private sector is involved in the provision of water services. Population growth in Guinea is high and urbanization is proceeding rapidly, putting strong pressures on service expansion. The urban population is growing at a rate of 5.8 percent a year, or twice the national population growth rate. One major concern is insufficient coordination between the public and the private sectors regarding investment requirements, risk allocation, and the tariff regime.

Improvements in the availability and quality of water service in Guinea have been made possible by a substantial increase in capital expenditure. During the first five years of the lease contract, the state-owned national water authority, SONEG, invested about $60 million in the water supply system. These resources were funded through a credit from the International Development Association and from internal cash generation. However, SONEG has had difficulty providing the capital needed for investment, effectively carrying out its regulatory functions, and monitoring compliance of the private contractor, SEEG, with contract obligations. In response to a substantial tariff increase, commercial losses have risen, as has the level of nonperforming consumer accounts. These circumstances have contributed to the erosion of SONEG's financial capacity.

SONEG bears financial responsibility for works that directly affect SEEG's operating results. This sharing of risks creates a situation in which no clear commercial incentives are brought to bear on the private management company in its operation and maintenance roles. "The efficient coordination of new investment with operations and maintenance will remain problematic as long as commercial risks are shared between the two entities, and, in particular, as long as SONEG remains as the principal financier of works that contribute to SEEG's effectiveness as an operator" (Brook-Cowen 1996).

Tailored Financial Strategies

The analysis thus far demonstrates that there are no miracles in financing investment in the water and wastewater sector. The notion that in the developing countries the financing challenge can be solved through private sector investment is a myth. This misconception might affect further developments in two ways. First, unrealistic expectations cannot be sustained over long periods. Discontent arising from disappointing outcomes can impair or even interrupt reforms that are in essentially the right direction. Second, sustained support of impractical strategies can contribute to institutional confusion in the sector, cause complementary options that are workable to be underestimated, and open a window for other entities, local and national governments, and financial institutions to elude their responsibilities.

Conclusions

Four major conclusions can be drawn from this analysis.

First, unless significant changes are introduced in the system of incentives, risks, and rewards facing the private sector, the bulk of financial resources for the water and wastewater sector in developing countries will come from the cash generated by the utilities themselves and from lending operations leveraged with the resulting cash flow. Cash generation is increasingly being paid for by consumers through tariffs. Thus, the ultimate guarantee to investors and lenders will depend on the effectiveness of politically and socioeconomically sustainable tariff regimes.

Second, investments have in general not proceeded according to expectations. There is a tendency in private sector arrangements to concentrate a large proportion of the investment program during the first phase of the contract. This puts strong pressures on tariff levels at the beginning of the arrangement. Greater coherence should be pursued among investment targets, tariffs, and quality standards.

Third, a coherent strategy for private sector participation in the water and wastewater sector should distinguish between two different approaches. The first of these is appropriate for water and sewerage systems that are already relatively well consolidated, with high coverage levels, particularly in growing and stable economies with developed capital markets and reliable institutions. In these circumstances private investors should be encouraged to take financial risks and compete for credits in financial markets to guarantee efficient management and operations, expansion, rehabilitation, and system maintenance.

The second approach is recommended for low-income economies with underdeveloped capital markets, low coverage levels, rapid population growth, increasing demand for expansion, and weak institutions. Here the rationale for private investment follows a different story. Private investors will tend to reduce their own risk, funding investment as much as possible out of cash generation. In a well-operated utility in a typical developing country, these resources usually cover operation and maintenance costs and part of capital expenditure but do not cover all investment needs for expansion and wastewater treatment.

Under the latter approach the public sector may have to maintain a financing role in the water and wastewater sector for some years to come. The principal challenge is to find the right mix between public and private finance. The role of private companies needs to be focused first on developing managerial and operating skills, improving the quality of services, increasing productive efficiency, formulating comprehensive investment plans, introducing accountability, and transforming unviable enterprises into financially viable companies capable of receiving and administering credits and government (or multilateral) funds. In this context, options other than concessions may be more suited to improving service delivery. The private sector can contribute to reducing—but will not eliminate—the need for government financing and the obligation on governments to develop financial mechanisms, strengthen capital markets, and provide guarantees and subsidies when necessary.

Fourth, the cases discussed in this study do not establish a correlation between the degree and type of private involvement and the magnitude or quality of the benefits obtained from private sector participation. In practice, concession arrangements do not necessarily guarantee better performance than do operation and management or lease contracts. From the perspective of optimizing investment, in an ideal world concessions may be preferable to lease or operation and management contracts, given the advantages of assigning responsibility for investments to a commercial entity that is also responsible for operations. However, concession arrangements are not always feasible, especially when investment costs exceed the borrowing capacity of the private sector, in countries where capital markets are weak, in small towns, or in cities with large low-income populations.

Although a combination of public and private finance may be needed in many cases, it is not free from difficulties. Bringing private operation and maintenance together with public finance—as was done in Cartagena, Gdansk, and Guinea, for example—has its drawbacks as well. Under such a scheme, private companies have few incentives to save on capital costs, and the efficiency of and accountability for investment often become a problem. An alternative that deserves further analysis is the design of partly financed concessions with an upfront public contribution assigned to the achievement of specific, strategically scheduled targets. Once the most expensive investment needs are met and system reliability improves, the private sector can assume a growing responsibility for subsequent investment programs. In the same way, after the system has been upgraded with publicly financed investments, management or lease contracts could be transformed into concessions, under which private firms would participate in mobilizing finance for future investments.

One major bottleneck for further expansion of private sector participation in the water and wastewater sector is the high concentration of firms on the supply side. Competition is restricted by the tendency for a few world players to dominate the sector. This is shown by the fact that a handful of international companies continuously reappear in the cases reviewed. For several reasons these large firms are not willing to participate in investment or in the provision of water service in many small towns and cities.

The design of policies that will encourage the entrance and development of local firms in the sector should therefore be a priority. Some encouraging evidence pointing to the incipient development of private local firms in the water business can be found in Chile, Mexico (Cancún, Aguascalientes), and Argentina (Corrientes province).

Guinea's experience reveals that scale remains another serious constraint to attracting private investment into small cities. Foreign companies must cope with the problems of paying and maintaining skilled personnel for long periods in small towns with few customers and poor infrastructure. For instance, Saur claims to need at least eight expatriates in Conakry to run the utility properly. With only 20,000 paying customers, and with low household incomes, this is hardly feasible. Without innovative practices and incentives to induce small municipalities to get together to contract with a private operator, for example, or other types of guarantees to induce the private sector to retain skilled personnel in small towns, private sector participation may be confined to capital cities and to mostly middle-income countries.

The development of local capital markets, the expansion of guarantees that encourage long-term private lending, and the development of other innovative financial instruments to provide accessible credit to investors and private operators remain important challenges for national governments and multilateral lending institutions as well.

Notes

1. Studies by the National Council of Public Works and the World Bank show that this ratio ranges from 3 to 4 in the case of telephone companies and electric utilities, is about 7 for toll roads, and is about 10 to 12 for water systems.

2. In the Buenos Aires concession, for example, a majority of the shares of Aguas Argentinas may not be pledged or transferred without the approval of

the Argentine government.

3. In the framework of the Colombia Water and Sewerage Sector Project (Loan 2961-CO), the World Bank currently participates in financing a water supply and sewerage subproject in Cartagena with a total cost of $55 million, of which about $26 million is financed by the Bank.

7
Regulation

The water and wastewater sector lends itself by nature to monopoly provision, making government intervention justifiable and perhaps inevitable. In the absence of intervention, monopoly suppliers of all types are tempted to charge excessive tariffs, provide poor service, or both.

In theory, regulation can improve economic efficiency by inducing natural monopolies to produce at a socially optimal level. When the regulatory framework is clear and stable, regulatory authorities strive to maintain a balance between the conflicting interests of governments, the regulated utilities, and consumers. In practice, however, such balance is difficult to achieve, let alone sustain. One of the dangers of responding to multiple and often opposing interests is that regulatory authorities may amend their decisions too often, creating an environment of uncertainty and distrust among investors and consumers.

Another concern is the quality of intervention. Regulators may tend to intervene excessively in the marketplace, for example by interfering with investment decisions and managerial control. Frequently regulatory agencies place too much emphasis on punitive threats, which are often ineffective, and too little on appropriate positive incentives and on enhancing the legitimacy of the regulatory process through equitable, just, and transparent policy decisions. A third problem has to do with the economic costs of regulation. As pointed out by Bradburd (1992), regulatory failure can foster efficiency losses greater than the deadweight monopoly losses it is intended to avoid.

Governments must find effective ways to design responsive regulatory policies and to identify the right mix of instruments and incentives to implement them. Whatever the model of private sector participation adopted, regulated companies and regulatory authorities face a steep learning curve in their attempts to define their relationship with each other and ensure an equitable and efficient provision of services.

Regulation takes place in a particular institutional environment, where the key factors are the type and quality of the incentives provided. The costs and benefits of regulation, therefore, must be estimated not on the basis of what the ideal institution would produce but on a realistic assessment of actual institutions and governmental effectiveness. Regulatory models and institutional arrangements cannot be easily transferred from one country to another. Instead regulatory systems need to be tailored to suit the specific economic and social distortions of the economies in which they will operate, their characteristic institutional endowments, and the effects of regulatory legislation on consumer habits.

The following critical regulatory issues need to be addressed and evaluated in all developing countries in which the private sector will participate in service provision, whatever the model:

- Definition of the roles and objectives of the regulatory entities
- Design of the regulatory entities and of the financial mechanisms that will allow them to fulfill their obligations
- Design of adequate incentives and regulatory instruments
- Development of technical expertise
- Choice of strategies to deal with informational asymmetries
- Enforcement of regulatory policies by institutions with limited means and capabilities
- Choice of strategies to maintain the independence of the regulatory agencies, to prevent their "capture" by experienced private companies or by political interests
- Choice of strategies to maintain the legitimacy and openness of the regulatory decision process.

This chapter will not attempt to analyze these issues in detail, but rather will describe in general terms the various ways in which governments and regulatory agencies in the six cases reviewed here have confronted these challenges. The chapter begins with an overview of the regulatory experience and the most critical tasks faced by regulators in Santiago and Buenos Aires. Then the focus shifts to the various rationales for regulation under the contracts in Cartagena, Gdansk, and Guinea. The chapter concludes by summarizing some other common barriers to and options for effective regulation in developing countries.

Lessons from Chile and Argentina

In Chile the institutional design and content of the regulatory framework for the water and wastewater sector were established before private sector participation was initiated. Water regulation benefited from the lessons learned from the privatizations in the electricity and telecommunications sectors initiated in the early 1980s. The Superintendency of Sanitary Services (SSS) was created in 1989 as an autonomous, decentralized entity, whose head is appointed by the president. The SSS depends administratively on the Ministry of Public Works; its staff consists of approximately 100 employees, and operational expenses are covered by the national budget. Its basic responsibilities are to grant concessions for water supply and sewerage services; define tariff regulation and control its implementation; and monitor compliance through norms and technical standards to ensure quality of services.

The regulatory experience in Chile highlights two major themes. First is the importance of putting in place clear incentives that are consistent with the country's macroeconomic policy and based on principles of economic efficiency. Second is the dynamic of the regulatory process and the need to develop innovative regulatory approaches to meet changing circumstances. The main characteristics of the Chilean regulatory regime are as follows:

- Clear separation of operating and regulatory functions and the definition of an appropriate set of incentives prescribed by top levels of government
- Implementation of a tariff structure based on efficiency criteria and easily understood by customers, with tariffs set for five-year periods on the basis of marginal cost for a "model enterprise"
- Incorporation of a coherent system of direct subsidies for low-income consumers
- Use of a long-term profitability concept under which water companies are expected to achieve a minimum rate of return.

Although in the Chilean regulatory system the public holding corporation, CORFO, does not formally regulate enterprises, in practice it plays a decisive role in the performance of regional public utilities. The ministers of planning, finance, and economy serve on CORFO's board of directors, providing consistency between sectoral and macroeconomic policies. As the largest shareholder in Chile's water companies, CORFO demands some minimum level of profitability and, in this way, exercises a regulatory role. CORFO is studying the idea of fully privatizing some companies whose performance has been poor or that have shown themselves unable to develop projects of inherent managerial complexity and/or large capital investments. If it is someday to regulate private companies, the SSS will have to be reinforced as privatization moves forward.

New regulatory approaches will be needed to deal with privately owned water companies. This is important because public regulation cannot pretend to regulate private companies effectively in the same way that it regulates public ones. Chile's experience in other infrastructure sectors shows that, although the regulatory system is relatively sophisticated, regulators have experienced serious difficulty in gathering adequate cost information from privatized utilities. Another emerging difficulty is the imbalance in negotiating power, given the increasing economic and political influence of private firms. Finally, as noted by Bitran and Serra (1995):

> . . . the limited information and technical capacity of regulators and the enormous influence of the private utilities in the rate-setting process have prevented the full transfer of efficiency increases to consumers via price reductions, even allowing for significant time lags. This situation has led to significant increases in the profits of regulated firms in the electricity and telephone services. Only in those cases where competition has emerged have drastic price reductions occurred.

The Argentine experience is instructive about the early stages of the regulatory process and the major challenges faced by an inexperienced public regulator in dealing with an experienced private operator. The Buenos Aires concession was the first large-scale example of private sector participation in the water and wastewater sector in Latin America (Box 7.1.). As in other such cases, the personnel of the regulatory entity were initially drawn from the former operating company. During the

Box 7.1. The Regulatory Scheme in Buenos Aires

The regulatory framework for the Buenos Aires water and sewerage concession became effective in April 1993. In contrast to that in Chile, the regulatory system in Buenos Aires proceeded in parallel with the transition from public to private operation. The regulatory framework defines the basic guidelines under which water and sanitation services are to be provided and prescribes the rights and obligations of the concessionaire (Aguas Argentinas), the Argentine government, and consumers with respect to the concession contract. The regulatory authority, Ente Tripartito de Obras y Servicios Sanitarios (ETOSS), is an autonomous entity in charge of monitoring Aguas Argentinas's performance and verifying fulfillment of its contractual obligations. ETOSS's main responsibilities include the revision and approval of expansion and maintenance projects, review of the tariff structure, and monitoring of Aguas Argentinas's compliance with investment plans.

ETOSS's board of directors consists of six members, with equal representation from the central government, the municipality of Buenos Aires, and the larger province of Buenos Aires. ETOSS is organized into six departments and has a staff of about 110. It is financed by consumers through a surcharge of 2.67 percent on the water and sewerage bill. This form of financing has the virtues of providing predictable revenue and ensuring a degree of regulatory autonomy from government interference. The challenge is how to set the right budgetary level, one that ensures the accomplishment of the prescribed regulatory functions without strangling the regulatory body or creating incentives for overintervention.

early years of the concession, the regulatory agency, ETOSS, had to focus its objectives on building regulatory capacity so as to level the playing field with the more experienced concessionaire. Relations between ETOSS and the concessionaire have been complex and tense.

According to the concessionaire, ETOSS's internal organization into six separate departments creates a problem of segregation of responsibilities and slow response times. One option to explore for remedying this problem is the appointment of a technical executive director within ETOSS, who would combine all technical, legal, and administrative matters in one office to facilitate more rapid and efficient decision making.

Regulatory rules can be thought of as incomplete contracts between regulators and the regulated companies, private or public. As a consequence, there will always be room for debate over the regulatory function and the interpretation of contracts and regulations. Buenos Aires has witnessed an interesting debate over these two issues. Whereas the concessionaire defends the character of the concession contract as one that specifies results rather than means of achieving them, the regulator believes that as the representative of the owner of the system—the Argentine government—it has the obligation to ensure that results are achieved with adequate procedures and high quality standards. Thus, the agency emphasizes that the contract must be understood as a contract of means as well as results. The key issue is the degree of freedom that Aguas Argentinas should have to fulfill the contract's targets through "investment optimization strategies," in other words, to achieve a given objective with less capital investment and with higher profits. ETOSS is concerned about the quality and the sustainability of investments. This problem will tend to become more acute during the later stages of the concession.

Availability of unmanipulated information remains one of the key problems facing regulatory agencies. Buenos Aires is no exception. According to ETOSS, the information the agency received from Aguas Argentinas during the first years of the concession was "poor, incomplete, and biased." For its part the concessionaire argues that the regulator requested excessive, unnecessary information with neither a clear purpose nor the necessary analytical capacity to process it. Procedures have improved—although there is room for further improvement—with the establishment of guidelines for providing the necessary information periodically using explicit reporting criteria.

The Experiences of Cartagena, Gdansk, and Guinea

The rationale for regulation of the private sector contracts in Cartagena and Gdansk is different from that in Buenos Aires and Santiago. In

Cartagena and Gdansk the local governments have regulatory responsibility both as shareholders of mixed companies and as contract regulators. This scheme poses an implicit conflict of interest to the extent that the government both represents the public interest and has an incentive to maximize the company's returns. But at the same time this model has the advantage of being self-regulating in the sense that the municipality and the operator are associates who have to reach key decisions by common agreement, rather than separate entities in confrontation with each other.

The regulatory experience in Colombia shows that, although the establishment of a well-designed regulatory framework is an important step, it is not enough to guarantee effective regulation and achieve the desired institutional changes. It is fundamental to understand why policies that outside observers perceive as optimal are not adopted, how institutions actually work, and the impact of this dynamic on the utility's performance. In July 1994 the Colombian Congress approved the Public Services Law (Law 142), which includes a regulatory framework conducive to expanding private sector participation, bringing about efficiency improvements, and promoting greater accountability of public service utilities. The law created a new Water Regulatory Commission (Comisión de Regulación de Agua Potable y Saneamiento, or CRA), placing it in charge of overseeing the water and wastewater sector, defining efficiency criteria, setting tariffs, and promoting competition for management of services. In addition, the government created a Superintendency of Domestic Public Services with responsibility for monitoring the performance of water companies, public as well as those with private sector participation, and enforcing regulations.

The impact of these reforms and the role of the CRA in the case of Cartagena have been marginal. Responsibilities for regulation are not clear. The operation and management contract assigns a wide range of regulatory and policy functions to the municipality. However, other central agencies and ministries also have regulatory functions, with fragmented and at times overlapping responsibilities. The tariff structure and tariff levels can be set by Acuacar under procedures and standards defined by the CRA. Even though the legislation clearly defines the objectives of the CRA, the agency is still struggling to find the appropriate mechanisms and instruments.

The CRA is funded by the public budget. The lack of a predictable mechanism of financing constitutes a weak point regarding the independence of its decisions. But the biggest challenge facing the CRA is how to implement effective regulation in a country with more than

1,020 separate public companies serving populations ranging from 5,000 to 6 million, and with diverse forms of ownership and levels of administrative and operational efficiency. Decentralizing regulation is an issue that merits more research in this type of institutional setting.

In countries with little or no history of private sector provision of public services, or in the case of small municipalities with weak administrative environments, greater emphasis on contracts and self-regulatory principles that minimize the costs of regulation can help. For example, the board of directors in Cartagena is composed of five members: two nominated by the municipality, two by the private operator, and one by the private shareholders. Decisions are reached on the basis of an 80 percent majority: at least four members must assent. This means that important decisions have to be fully supported by both the municipality and the private operator.

However, reliance on contract provisions in the hope that the need for regulation can be reduced has proved unrealistic. For example, the concession contract in Buenos Aires was renegotiated after only one year to incorporate revised priorities for system expansion. An additional drawback of the quest for the "perfect" contract is the expensive delays caused by protracted negotiations. A reasonable degree of contract flexibility is likely to be necessary. The dynamics of the market after entry of the private sector participant make initial financial, economic, and institutional forecasts unreliable. This does not, however, argue for leaving everything open; it is essential to specify which procedures will be used in the event discrepancies appear.

Lease contracts and operation and management arrangements also require a high level of administrative capacity and sound political support for contract enforcement. Contracts of this type are not necessarily easier to administer and regulate than contracts for more invasive private sector involvement such as concessions:

> In Guinea, the government has had difficulty monitoring SEEG's compliance with the contract, and, thus in effectively carrying out its regulatory functions with regard to the tariff received by SEEG. SONEG appears to have given greater weight to its investment functions than to rigorous pursuit of its monitoring functions. At the same time, SEEG appears to have been slow in complying with financial reporting requirements. One factor that may explain monitoring and enforcement problems is the fact that the government has played a passive role as co-owner of SEEG, rather than using its presence on SEEG's board to rigorously monitor performance and enforce compliance with reporting requirements. (Brook-Cowen 1996, p. 4)

Other Regulatory Issues: The "Independence" Debate

Experience shows that in developing countries "independent" regulatory capacity is an elusive goal. The water and wastewater sector has been and remains subject to strong political influence. If politics is the art of defining who gets what, regulation is in essence a political exercise. In many respects, with the advent of private sector participation the *form* of government intervention in the water sector has changed, as the sector becomes increasingly subject to more comprehensive and refined types of intervention. However, what is really important is not whether a regulatory entity is "independent" or not, but whether a government can guarantee, formalize, and institutionalize its commitments to investors and consumers. What governments usually do to show that they are able to guarantee fulfillment of certain basic rules is establish a regulatory framework that *limits* discretion in pricing; one way to do this is by creating commissions to protect the regulatory decision-making process from short-term political interests. Montenegro (1995) describes how the regulatory commissions in Colombia were created:

> In Colombia the Government's initial proposal was to establish fully independent commissions [for the telecommunications, electricity and water supply sectors]. Some ministers and members of Congress insisted on, and eventually propelled, the creation of more public and thus less autonomous entities. Some ministers argued, for example, that tariff-setting could have a strong political impact and hence that responsibility should not be left in the hands of technical experts, since the government would be blamed when their decisions turned out poorly. Likewise, some members of Congress feared that regulations crafted solely by experts would be excessively technocratic, without regional and consumer input. Therefore Congress compromised by approving the above mentioned three semi-independent regulatory commissions.

Some of the safeguards most often recommended to promote regulatory independence include fixed terms for regulators, to give them security of tenure; providing regulatory boards with an independent source of funding, to prevent politics from driving budgetary allocations; and prescribing high professional qualifications for appointment. A related point is the risk of manipulation of regulatory decisions by relatively powerful private companies. In developing countries private conglomerates and other special interest groups not infrequently exert pressure in an effort to circumvent top executive or legislative authorities and countervail "benign" regulations or anticompetitive practices. High standards of transparency and strict information disclosure can limit the incidence of this problem.

The development of antitrust regulatory policy will become increasingly important as private sector arrangements become more widespread. Unfair competition can arise when any water company splits up its business into subsidiary companies or "facilitates" bids and contracts for related companies (for example, construction companies, pipe producers, meter manufacturers) and enters into subcontracting arrangements with them. In Buenos Aires, for instance, some local professional associations have claimed that approximately 70 percent of expansion works were contracted with companies associated with Aguas Argentinas's shareholders, while the remaining 30 percent were awarded to French subsidiaries of the Dumez conglomerate.

Decentralization of central government functions presents opportunities but also carries risks and new challenges for municipalities. Many countries have transferred responsibilities for public service provision to local governments. The paradox of decentralization is that it seeks to reduce dependence on central governments but at the same time, given the financial constraints and the institutional weakness of municipalities, requires strong support at the national level. Most municipalities in developing countries lack the capacity to deal with the experienced private companies they must regulate on equal terms. National principles and guidelines combined with regional or local instruments for resolution of specific conflicts seem an appropriate approach. There is also an optimum scale for efficient regulation. Regulating a small number of companies might prove ineffective because the regulators lack sufficient information for performance comparison. Regulating a large number of firms, however, might prove too costly.

Some (for example, Bradburd 1992) have argued that at least in the initial stages of private sector participation—given the high costs of regulation, economic distortions, and weak institutional capacity characteristic of the developing world—the main benefits of regulation should not be expected to arise from minimizing the efficiency losses due to monopoly power. Instead they will flow from gains associated with other key objectives of monopoly regulation, especially those involving income distribution and promoting credibility and investor confidence in institutional capacity and stability.

Lessons

The cases reviewed illustrate the following lessons concerning regulation:

- Regulation in any given country takes place in a unique institutional context, where the key issues are the type and quality of incentives provided. Regulatory models and institutional arrangements cannot easily be transferred from one country to another. Regulatory costs and benefits, therefore, must be estimated not on the basis of ideal institutions but on a realistic assessment of actual institutions and government effectiveness.

- The creation of a regulatory framework does not guarantee effective regulation or the achievement of desired institutional changes. It is fundamental to understand why governments do not always adopt policies that outside observers recommend, how institutions work, and the impact of this dynamic on performance.
- There is an imbalance between the limited means and capabilities of public regulators in developing countries and the capacity of experienced private operators. Public administration will need to be strengthened to enforce regulatory regimes and provide guarantees that improve the institutional and financial viability of projects in which the private sector participates.
- Further private sector participation may be impeded by the high concentration of water companies on the supply side, with competition restricted to a few world players. Policies will need to be designed to encourage the entrance and development of local firms in the water supply sector.

8
Lessons and Recommendations

The participation of private enterprises in the provision of water and wastewater services has great potential for improving the efficiency and quality of service delivery in developing countries. However, experience testifies that the development of private sector arrangements is not free of risks and difficulties. This chapter summarizes the principles and lessons that can be drawn from the cases reviewed in this study.

Quantity and Quality of Service Provision

Private sector participation has led to substantial benefits to consumers in terms of expanded coverage and quality of services as well as significant improvements in productive efficiency. However, consolidating these gains and reaping additional benefits in the future will depend heavily on strong leadership and continuous political commitment, as well as on the ability of governments and financial institutions to implement complementary reforms, especially in the areas of water pricing, financing, and regulation.

Efficient Pricing of Water

Despite a growing consensus on the need for more efficient tariff systems that give the right signals to investors and consumers, many barriers have impaired the implementation of such systems. Most privatized utilities have inherited or adopted inefficient tariff structures.

The evidence reveals a tendency to set unrealistic goals, especially in the initial stages of private sector participation. Investment plans should be balanced more consistently against potential revenues from tariffs. Greater attention should be dedicated to assessing the impact of metering and full-cost pricing on consumption.

Tariff structures that embody highly distortive cross-subsidies continue to create disincentives for expansion of services to the poor. Private sector providers must be given positive incentives to serve the poor, who in turn must be helped to pay for the cost of service through targeted subsidies financed from government budgets.

Evidence shows that tariff setting continues to be influenced strongly by political factors after the private sector service provider enters the picture. The lack of clear policies and procedures for tariff adjustments invites such influence. The methodology for calculating tariffs should be transparent and should specify objective criteria, which when met should automatically trigger tariff adjustments.

Innovative Financial Strategies

The financial challenge in the water and wastewater sector remains. Evidence indicates that, given the current balance of risks, incentives, and rewards, the private sector is unlikely to invest its own resources in the sector in developing countries. Innovative financial instruments and more effective incentives need to be introduced to induce further private sector involvement. In the medium term the bulk of the sector's financial resources in developing countries will come from cash generation and from lending operations leveraged with this cash flow.

The success of future financing of private investment in developing countries' water and wastewater sectors will hinge on the design of comprehensive long-term financial strategies. The private sector can play an important role here. But this role is nevertheless a limited one and should be managed in a more pragmatic and balanced way, to respond to specific challenges, risks, and opportunities.

The appropriate role of the private sector in water and wastewater service provision will differ according to its setting. In countries that have relatively consolidated water and sewerage systems, with high coverage levels, developed capital markets, and reliable institutions, private investors should be encouraged to take financial risks and compete for credit in the financial markets. Doing so will promote efficient management and operations, expansion, rehabilitation, and maintenance of systems. On the other hand, in low-income economies with underdeveloped capital markets, low coverage levels, increasing demand for expansion, and weak institutions, private investors can contribute to reducing—but cannot eliminate—the need for government financing. In such countries the principal challenge is to find the right mix of public and private finance. The strategies of private companies will need to be focused on developing managerial and operating skills, increasing productive efficiency, formulating comprehensive investment plans, and increasing creditworthiness.

The availability of long-term financing at reasonable interest rates remains the critical constraint on further private sector investments in the water and wastewater sector. The development of local capital markets, the expansion of loan guarantees that encourage long-term private lending, and the development of other innovative financial instruments to provide credit to investors and private operators remain major challenges for national governments and multilateral lending institutions.

Concession arrangements do not necessarily guarantee better performance than do operation and management arrangements or lease contracts. From an investment viewpoint, concessions may be the preferred approach, given the advantages of assigning responsibility for investment to a commercial entity that will also be responsible for operating the facilities. However, concession arrangements are not always feasible, especially when the cost of investment exceeds the borrowing capacity of the private sector, in countries where capital markets are weak, or in towns and cities where a large proportion of the population has low incomes.

Improving Regulation

Effective regulation is the cornerstone of sustainable private sector participation. Yet there are no universally applicable regulatory models or institutional arrangements that can be simply transferred from one country to another. Regulation takes place in a particular institutional environment, where the determinant variables are the type and quality of incentives provided.

Successful regulatory reform will require effective institutional reform as well as stronger public administration, to enforce regulatory regimes and provide guarantees to investors and consumers that improve the institutional and financial viability of projects in which the private sector is involved.

A tremendous imbalance exists between the limited means and capabilities of public regulators and the capacity of experienced private operators. Some of the more critical regulatory issues that will need to be addressed, regardless of the model of private sector participation, include the definition of regulatory functions and interpretation of contracts; the setting of appropriate incentives; and strategies to deal with informational asymmetries, with the threat of capture of the regulatory agencies by the regulated companies or by political interests, and to increase the perceived legitimacy of regulatory decision making.

Establishing a sound regulatory framework is an important first step, but it does not guarantee effective regulation or the achievement of desired institutional changes. It is fundamental to understand why recommended policies are sometimes not adopted, how institutions work, and what the impact of this dynamic is on utility performance.

One major bottleneck to further participation of private firms in the water and wastewater sector is the high concentration of these firms on the supply side. Competition is limited because the sector is dominated by a few world players. Moreover, this situation has contributed to a tendency to concentrate private sector initiatives in large urban centers. This hampers the design of long-run institutional and financial solutions for smaller cities and towns. To improve the quality of procurement, promote competition, and facilitate regulation, priority should be given to designing policies that will encourage the entry and development of local firms.

It is important to reduce the excessive transaction costs and delays so often incurred in designing and awarding contracts. International lending organizations such as the World Bank could contribute to this objective by preparing indicative model contracts and bidding documents to which national and local governments can refer; by providing timely advice from independent consultants; by designing training programs for local officials, consultants, and domestic firms in specific areas of expertise; and by designing systematic monitoring mechanisms to assess the results, impact, and difficulties of projects undertaken and to draw lessons for improving the quality of these and other projects, present and future.

Private sector involvement in public service provision would also benefit from better coordination among the multilateral lending organizations and their affiliate agencies, the regional development banks, and other financial institutions. Better dissemination of updated information between them and their clients is essential to improving projects that involve private sector participation.

The issue of minimum size—how large the scale of a project must be if it is to attract private investment—constitutes a serious constraint on private sector participation in the public utilities of small cities. Unless innovative practices and incentives are devised to promote joint contracting by small municipalities, or other types of guarantees that will induce the private sector to maintain skilled personnel in small towns, private sector participation may be confined to capital cities and mostly middle-income countries.

Areas for Further Research

The experiences with private sector participation reviewed in this study are few and recent. Their political, socioeconomic, and other characteristics will not necessarily match those in other countries and cities. Although analysis of these experiences can offer valuable lessons for other cities and countries and suggest basic principles and guidelines, it may not be possible to apply similar strategies in the same way elsewhere. Further analysis is needed of these six cases and of other experiences with private sector participation, on the assumption that the pursuit of reform in the water and wastewater sector will lead to improvements in the quality of life for all segments of society, especially the poor.

Some issues that merit further study are the exploration of viable alternatives, including improvements in the design of direct subsidies, to provide the poor with better access to services; the fiscal and socioeconomic impact of private sector participation; risk analysis and expansion of guarantees for investors; regulatory capacity building at the local and the national level; the design of innovative regulatory tools such as performance-based regulation, economic instruments, and self-regulatory mechanisms; guidelines for negotiating staff retirement programs; and promotion of private microenterprises among the former employees of utilities that have experienced work force reductions.

Appendix. Principal Features of Private Sector Arrangements in the

Feature	*Buenos Aires*	*Cancún*	*Cartagena*
Type of arrangement	30-year concession	30-year concession	26-year operation and maintenance contract
Date of transfer	May 1993	January 1994	June 1995
Private sector participant	Aguas Argentinas (consortium of foreign and local investors and IFC)	Grupo Mexicano de Desarrollo through operator AGUAKAN	ACUACAR (mixed-capital company 50% owned by municipality, 45% by Aguas de Barcelona)
Service area	Capital city and 14 surrounding districts	Municipalities of Cancún and Isla Mujeres	District of Cartagena
Population served	8,600,000	337,000	750,000
GNP per capita (1994 dollars)	8,060	4,010	1,620
Coverage before private sector participation	Water 70%, sewerage 58%	Water 61%, sewerage 34%	Water 70%, sewerage 58%
Targets	Water coverage 100%, sewage treatment 93%, UFW reduced from 43% to 25%, construction of sewage treatment plant	Water coverage 95% (by 1996), sewerage coverage 95% (by 1998), sewage treatment 100% (by 1996)	UFW reduced from 52% to 25% and collection rate raised from 62% to 100% in 10 years; improved quality standards and maintenance
Investment requirements	$4 billion (of which $1.2 billion in first 5 years); backlog exists in wastewater treatment investments	$120 million over 1994–2000	$250 million over next 5 years

Note: IDA, International Development Association; IFC, International Finance Corporation; UFW, unaccounted-for water.

Water Supply and Wastewater Sector

Gdansk	*Guinea*	*Santiago*	*Corrientes*[a]
30-year operation and maintenance contract	10-year lease contract	Service contracts	30-year concession
December 1992	August 1989	1990	September 1991
SNG (mixed-capital company 51% owned by foreign investor, 49% by municipality)	SEEG (mixed-capital company 51% owned by foreign investors, 49% by state)	EMOS (public stock corporation 64% owned by state development corporation, 35% by central government)	Aguas de Corrientes (jointly owned by local and foreign investors)
Cities of Gdansk and Sopot	Capital city (Conakry) and 16 other towns	Capital city and 21 periurban localities	10 principal cities of the state
500,000	1,500,000	5,000,000	520,000
2,470	510	3,560	8,060
Water 96%, sewerage 96%	Water 40%, sewerage data not available	Water 99%, sewerage 91%	Water 62%, sewerage 30%
Improved quality of water and of effluents; improved maintenance and rehabilitation of network	Improved service and performance standards; maintenance and rehabilitation responsibilities	Improved quality of services and standards	Water coverage 100%, sewerage coverage 75% in 15 years; improved quality standards
$100 million over next 5 years	Not available	$600 million for wastewater treatment over next 15 years	$75 million over next 15 years
$4 million per year			

(Table continues on the following pages.)

a. The data on Corrientes were based based on a brief review. Thus Corrientes was not included as a case study, but the information is included here for the benefit of the reader.

Appendix (continued)

Feature	*Buenos Aires*	*Cancún*	*Cartagena*
Financing	$98 million from IDB, $385 million from IFC, $330 million from cash flow, $120 million from equity	$10 million from cash flow, $12 million from concessionaire; negotiations under way with IFC	Cash generation; $2 million per year from Colombian government; discussions under way over $140 million World Bank loan
Tariff regime	Rates set by bid and raised 13.6% after first year. Adjustments to occur every 5 years. Current rates remain 17% below level under public operation.	Fixed charge plus per-unit charge that rises with consumption. No systematic adjustment instrument.	Fixed charge plus per-unit charge that rises with consumption from $0.12 to $0.86/m^3. No systematic adjustment instrument.
Metering	Fewer than 10% of residential users metered. Goal is to add 80,000 meters to system per year. 140,000 meters currently installed in industrial sector.	Coverage increased from 54% to 86% over 2 years; 16,000 meters currently installed.	Coverage increased from 56% to 61% over 6 months.
Subsidy mechanisms	Cross-subsidization allowed	Cross-subsidization between hotel and residential sectors	Cross-subsidization among 6 income groups
Regulation	Autonomous	Former state operator provides regulation and has encountered difficulties in monitoring and enforcement. Financing not well defined.	National regulatory commission. Financed by transfers from national budget. Regulatory responsibilities not clearly assigned.

Gdansk	*Guinea*	*Santiago*	*Corrientes*[a]
$4 million per year from cash flow; additional funding from municipality	Cash flow and $58 million IDA credit	Cash flow and $55 million World Bank loan	Problematic. Main sources are working capital and credits; $18 million allocated for sewerage plants.
Two user categories. Tariffs negotiated annually with city council. No systematic adjustment instrument. No clear mechanism for setting tariffs.	Tariffs set to cover operating costs, debt service, and return on assets. Tariffs increased from $0.24 to $0.90/m^3$. 10,000 connections (33%) inactive.	Tariffs based on efficiency criterion and long-run marginal cost for a "model enterprise."	Basic single tariff with price cap.
Metering is not by household but by "community." Coverage increased by 12% over 3 years.	Coverage increased from 5% to 95%.	Coverage is 100%.	Coverage increased from 2% to almost 100%; 85,000 meters currently installed.
None	Only in initial phase: difference between tariff revenues and operating costs funded by IDA credit on declining basis	Direct subsidy from central government budget for first 20 m^3 per month; annual subsidies total $25 million and benefit 450,000 families.	None
Regulation by city through municipal council. Process is complex and unclear. Political interference.	Regulation by state-owned water authority has encountered monitoring and enforcement difficulties. Unclear assignment of risks. Incentives for efficiency are weakened.	Regulatory agency with staff of 100 under Ministry of Public Works. Financed from national budget.	Regulatory agency with staff of 19 under Ministry of Public Works. Agency regulates and provides services. Weak regulation and political interference.

Bibliography

The word "processed" describes informally reproduced works that may not be commonly available through library systems.

Aguas Argentinas. 1993. "Contrato de Concesión entre el Gobierno Argentino y la Sociedad Aguas Argentinas." Buenos Aires. Processed.

————. 1994a. "Estados Contables al 31 de Marzo de 1994 Conjuntamente con el Informe del Auditor e Informe de la Comisión Fiscalizadora." Buenos Aires. Processed.

————. 1994b. "Memoria y Balance. Estados Contables al 31 de Diciembre de 1994 y 1993." Buenos Aires. Processed.

————. 1995. "Plan a Mediano Plazo 1995–1998." Buenos Aires. Processed.

Aguas de Cancún, S.A. (AGUAKAN). 1993. "Titulo de Concesión entre el Gobierno del Estado de Quintana Roo y los Municipios de Benito Juárez e Isla Mujeres en Favor de Desarrollos Hidráulicos de Cancún S.A." Chetumal, Mexico. Processed.

————. 1995. "Sistema de Agua Potable, Alcantarillado Sanitario y Saneamiento. Cancún, Quintana Roo. Informe de Evaluación de la Concesión." Cancún, Mexico. Processed.

Aguas de Cartagena (ACUACAR). 1995a. "Estatutos para la Constitución de la Empresa de Economía Mixta Aguas de Cartagena S.A." Cartagena de Indias, D.T. Processed.

————. 1995b. "Minuta del Contrato para la Gestión Integral de los Servicios de Acueducto y Alcantarillado." Gerencia del Programa de Acueducto y Alcantarillado. Cartagena de Indias, D.T. Processed.

Aguas de Cartagena (ACUACAR) and Price Waterhouse. 1996. "Balance General al 31 de Diciembre de 1995." Cartagena de Indias, D.T. Processed.

Alfaro, Raquel. 1995a. "Comentarios al Documento 'Análisis Comparativo de la Gestión de Empresas de Servicios Sanitarios y de Distribución Eléctrica' (Elaborado por el Instituto Libertad y Desarrollo)." Santiago de Chile. Processed.

————. 1995b. "Competition and Social Extension in a Water Supply and Sanitation Company." Santiago de Chile. Processed.

Barr, Nicholas, ed. 1994. *Labor Markets and Social Policy in Central and Eastern Europe: The Transition and Beyond.* New York: Oxford University Press.

Bitran, Eduardo, and Pablo Serra. 1995. "Regulation of Privatised Utilities: Lessons from the Chilean Experience." OECD Advisory Group on Privatization, Eighth Plenary Session, Paris, October 30 and 31. Paris: OECD. Processed.

Boland, John. 1995. "Expansion of User-Fee Based Systems." Paper presented at a World Bank training seminar on "The Enabling Environment for the Urban Water and Sanitation Sector," Baltimore, Md. Processed.

Bradburd, Ralph. 1992. "Privatization of Natural Monopoly Public Enterprises. The Regulation Issue." Policy Research Working Paper 862. World Bank, Public Sector Management and Private Sector Development, Washington, D.C. Processed.

Brook-Cowen, Penelope. 1996. "Private Sector Participation in Water Supply in Low-Income Countries: Lessons from Guinea at the 5-Year Mark." World Bank, Transport, Water, and Urban Development Department, Washington, D.C. Processed.

Canavy, Guy, and Fernando Porta. 1995. "Evaluación de la Primera Etapa de Administración Privada: El Ejemplo de Buenos Aires." Paper presented at a seminar on "Participación Privada en la Gestión de Aguas," Barcelona, October. Processed.

Cestti, Rita, Guillermo Yepes, and Augusta Dianderas. 1996. "Managing Demand for Urban Water Utilities." World Bank, Transport, Water, and Urban Development Department, Washington, D.C. Processed.

Comisión Nacional del Agua. Mexico. 1996. "Saneamiento Integral de Cancún, Quintana Roo. Diagnóstico, Inversiones y Responsables." Chetumal, Mexico. Processed.

Edwards, Sebastian. 1995. *Crisis and Reform in Latin America: From Despair to Hope*. New York: Oxford University Press.

EMOS (Empresa Metropolitana de Obras Sanitarios). 1995. "Memoria y Balance Anual 1994." Santiago de Chile. Processed.

———. 1996. "Para una Mejor Calidad de Vida." Santiago de Chile. Processed.

ETOSS (Ente Tripartito de Obras y Servicios Sanitarios). 1994. "Memoria y Balance 1993." Buenos Aires. Processed.

Hemming, Richard, and Ali M. Mansoor. 1988. "Privatization and Public Enterprises." Occasional Paper 56. International Monetary Fund, Washington, D.C. Processed.

Hervas, Jose Luis. 1996. "Desempeño de la Empresa de Aguas de Murcia: Su Transformación en una Empresa de Economía Mixta." In *Economia del Agua y Sector Privado*. Ministerio de Desarrollo Económico de Colombia, Santafé de Bogotá, D.C.

Idelovitch, Emanuel, and Klas Ringskog. 1995. *Private Sector Participation in Water Supply and Sanitation in Latin America*. Washington, D.C.: World Bank.

Libertad y Desarrollo. 1995. *Análisis Comparativo de la Gestión de Empresas de Servicios Sanitarios y de Distribución Eléctrica*. Serie Opinión Económica. Santiago de Chile.

Maksymiuk, Zbigniew. 1994. "Transformation de L'Entreprise Communale en Societe Joint-Venture: L'Example de la Société Anonyme Saur Neptun Gdansk." Gdansk. Processed.

Menendez Espino, José. "La Transición de la Empresa de Aguas de Murcia de una Empresa Pública a una Empresa de Economía Mixta." In *Economía del Agua y Sector Privado*. Ministerio de Desarrollo Económico de Colombia, Santafé de Bogotá.

Montenegro, Armando. 1995. "Economic Reforms in Colombia: Regulation and Deregulation, 1990–94." World Bank, Economic Development Institute, Washington, D.C. Processed.

Munasinghe, Mohan. 1992. *Water Supply and Environmental Management: Developing World Applications*. Boulder, Colo.: Westview Press.

North, Douglass, and Alberto de Capitani. 1994. "Institutional Development in Third World Countries: The Role of the World Bank." HRO Working Papers. World Bank, Human Resources Development and Operations Policy, Washington, D.C. Processed.

Palominos, Maria Concepción. 1994. "La Regulación y el Sistema Tarifario de los Servicios de Agua Potable y Alcantarillado en Chile." Paper presented at a seminar on "Abastecimientos de Agua," Murcia, Spain. Processed.

Paniza, Ricardo. 1995. "Empresas Mixtas en Cartagena de Indias." Paper presented at a seminar on "Participación Privada en la Gestión de Aguas," Barcelona, October. Processed.

Rais, Jorge. 1995. "La Regulación de los Servicios de Agua Potable y Alcantarillado en Buenos Aires." Paper presented at a seminar on "Participación Privada en la Gestión de Aguas," Barcelona, October. Processed.

Rees, Judith. 1995. "Protecting the Consumer." Paper presented at a World Bank training seminar on "The Enabling Environment for the Urban Water and Sanitation Sector," Baltimore, Md., September. Processed.

Saur. 1994. "Study of Gdansk." In "Private Management of Public Services," a presentation to the World Bank. Washington, D.C. Processed.

Serageldin, Ismail. 1994. *Water Supply, Sanitation and Environmental Sustainability. The Financing Challenge*. Washington, D.C: World Bank.

Triche, Thelma A. 1990. "Private Participation in the Delivery of Guinea's Water Supply Services." Policy Research Working Paper 477. World Bank, Transport, Water, and Urban Development Department, Washington, D.C. Processed.

———. 1995. *The Regulation of Urban Water Supply and Sewerage Services: Proceedings of the Seminar "Regulation of Utilities and Monopolies."* Washington, D.C.: World Bank, Economic Development Institute, and Institute of Public-Private Partnership.

Triche, Thelma, and Barbara Richard. 1994. "Reducing Regulatory Barriers to Private Sector Participation in Latin America's Water and Sanitation Services." Policy Research Working Paper 1322. World Bank, Transport, Water and Urban Development Department, Washington, D.C. Processed.

Van de Walle, Nicolas. 1989. "Privatization in Developing Countries: A Review of the Issues." *World Development* 17(5): 601–15.

World Bank. 1994a. *Poverty in Colombia*. A World Bank Country Study. Washington, D.C.

———. 1994b. *World Development Report 1994: Infrastructure for Development*. New York: Oxford University Press.

———. 1995a. *Meeting the Infrastructure Challenge in Latin America and the Caribbean*. Washington, D.C.

———. 1995b. *Understanding Poverty in Poland*. A World Bank Country Study. Washington, D.C.

———. 1995c. *The World Bank Atlas 1996*. Washington, D.C.

Yepes, Guillermo, and Augusta Dianderas. 1996. "Performance Indicators. Water and Sanitation Utilities. 2nd edition. Financial Indicators and Overview of Service Rates." World Bank, Transport, Water, and Urban Development Department, Washington, D.C. Processed.

Directions in Development

Begun in 1994, this series contains short essays, written for a general audience, often to summarize published or forthcoming books or to highlight current development issues.

Africa's Management in the 1990s and Beyond: Reconciling Indigenous and Transplanted Institutions

Boom, Crisis, and Adjustment: A Summary

Building Human Capital for Better Lives

Early Child Development: Investing in the Future

Financing Health Care in Sub-Saharan Africa through User Fees and Insurance

Global Capital Supply and Demand: Is There Enough to Go Around?

Implementing Projects for the Poor: What Has Been Learned?

Improving Early Childhood Development: An Integrated Program for the Philippines

India's Family Welfare Program: Moving to a Reproductive and Child Health Approach (with a separate supplement)

Investing in People: The World Bank in Action

Managing Commodity Booms — and Busts

Meeting the Infrastructure Challenge in Latin America and the Caribbean

MIGA: The First Five Years and Future Challenges

Nurturing Development: Aid and Cooperation in Today's Changing World

Nutrition in Zimbabwe: An Update

Poverty Reduction in South Asia: Promoting Participation of the Poor

Private and Public Initiatives: Working Together for Health and Education

Private Sector Participation in Water Supply and Sanitation in Latin America

Reversing the Spiral: The Population, Agriculture, and Environment Nexus in Sub-Saharan Africa (with a separate supplement)

A Strategy for Managing Water in the Middle East and North Africa

Directions in Development (*continued*)

Taxing Bads by Taxing Goods: Pollution Control with Presumptive Charges

Toward Sustainable Management of Water Resources

Trade Performance and Policy in the New Independent States

The Transition from War to Peace in Sub-Saharan Africa

Unshackling the Private Sector: A Latin American Story

The Uruguay Round: Widening and Deepening the World Trading System